How To Mine and Prospect for Placer Gold

By J. M. West

UNITED STATES DEPARTMENT OF THE INTERIOR
Rogers C. B. Morton, Secretary

BUREAU OF MINES
Elburt F. Osborn. Director

Fredonia Books
Amsterdam, The Netherlands

How to Mine and Prospect for Placer Gold

by
J. M. West

for U.S. Department of the Interior

ISBN: 1-4101-0893-7

Copyright © 2005 by Fredonia Books

Reprinted from the 1971 edition

Fredonia Books
Amsterdam, The Netherlands
http://www.fredoniabooks.com

CONTENTS

ILLUSTRATIONS

HOW TO MINE AND PROSPECT FOR PLACER GOLD

by

J. M. West [1]

ABSTRACT

Increased leisure time and increased interest in the out-of-doors is lead-
ing more and more families to experiment with placer mining of gold, and sometimes even to going on into small-scale production. This
report supplies basic information on areas of occurrence, equipment needed,
prospecting, sampling, mining, and regulations concerning the possession and
sale of gold. Selected references are given for further study.

INTRODUCTION

Placer gold has tantalized many a person who has tried his luck and skill
in the hope of striking it rich. Separating gold from embedded materials is
basically simple, and can be done effectively on nearly any scale, depending
upon the deposit and the capital available for investment. The final product
is consistently in demand at a relatively stable price. Historically, however,
one must be advised that rewards for the majority of small-scale miners--those
who operate "on a shoestring"--have been depressingly small.

First of all, the placer miner must know where placer deposits are located
and he must have the technical knowledge to extract the gold. Additionally,
he must face problems of land ownership, water supply, and water pollution,
all of which have grown in complexity with the population. The costs of labor
and equipment are relatively high now, although this may not seem significant
to an individual mining a small deposit. Secondhand equipment may become
available at relatively low cost because of a slowdown in construction or as
surplus at the end of a war. By taking advantage of such opportunities, one
can sometimes make an otherwise unprofitable operation successful, at least as
long as the equipment holds up.

To the novice or "weekend prospector," the more complex features of
placer mining may seem hard to comprehend. At any rate, the novice is often
more interested in the recreational values offered by gold placering than in
its profitability. Thus, the search for and discovery of even a small grain

---

[1] Physical scientist, Division of Nonferrous Metals, Bureau of Mines,
Washington, D.C.

or nugget of gold is an achievement worth considerable effort. As a start, the beginner may gain some benefit from visiting one of the many pan-for-a-fee tourist establishments typically found in gold-mining areas.

The small-scale miner may sell his gold, but often he keeps it as a souvenir, or for use in some kind of jewelry, or in the hope that its value will appreciate. Seldom is a placer gold venture truly profitable when all costs are considered under existing circumstances. On the other hand, an individual or a family can gain a great deal of pleasure and satisfaction from the experience of producing your own gold. Producing your own, even on a small scale, involves a number of problems which this publication attempts to discuss. Because the subject is so extensive, the reader is referred to other reports in the bibliography for more detailed information.

ACKNOWLEDGMENT

For many years, three Bureau of Mines Information Circulars written by E. D. Gardner and C. H. Johnson in 1934 and 1935 have been a basic reference on gold placering. However, it was realized after several reprintings of the initial volume of the series that the general presentation had become dated and often went beyond the scope of the usual requests for information. This report has borrowed heavily from the Gardner and Johnson material because of its adaptability, and the author wishes to acknowledge that source in particular, although many other sources have been used in preparation.

HISTORY OF PLACER GOLD MINING

Placer gold mining in the United States spans a period of nearly 200 years Earliest mining took place in the Eastern States and particularly in the southern Appalachian region during the late 1700's and early 1800's, but the richer deposits were soon exhausted, and interest turned to the West. The earliest production of any note in the West was from the Old and New Placer Diggings near Golden, Santa Fe County, N. Mex. These deposits were worked as early as 1828. A few other deposits were mined in the succeeding years until the first discovery of major importance, that of James Marshall on January 24, 1848, on the American River at Coloma, Calif. This discovery was a major factor in the rapid settlement of the West and triggered the first of the great gold rushes in the United States. Because of the lure and excitement of gold mining, prospectors spread throughout the West and in subsequent years many more rich placer gold deposits were found. A selected listing of discoveries subsequent to Coloma follows:

```
1848-49.... California.. Trinity and Klamath Rivers.
1849....... Nevada...... Gold Canyon.
1852....... Oregon...... Grants Pass district.
            Montana..... Gold Creek.
1857....... Nevada...... Six-Mile Creek.
1858....... Arizona..... Gila City.
            Colorado.... Cherry Creek, Ralston Creek, Platte River.
1858-60.... Washington.. Blewett Pass (northern and central parts of State).
1859....... Colorado.... Clear Creek, Blue River, Arkansas River.
1860-61.... Idaho....... Clearwater River, Pierce City, Oro Fino, Elk City,
                              Florence, Warren.
```

```
1862....... Montana..... Bannack, Alder Creek.
            Idaho....... Boise Basin.
            Arizona..... La Paz district.
1863-64.... Arizona..... Weaver Creek, Lynch Creek.
            Utah........ Bingham Canyon.
1864....... Montana..... Helena.
1867....... Nevada...... Tuscarora district.
            New Mexico.. Elizabethtown district.
1874-75.... South Dakota Black Hills, Deadwood Gulch.
1876-77.... Nevada...... Copper Mountain (Charleston district), Osceola.
1881....... Nevada...... Spring Valley.
```

In Alaska, gold occurrences were reported as early as 1848, and gold was found in the Yukon region about 1878; but not until the fabulously rich finds in 1897-98 in the Yukon's Klondike (in Canadian territory) did placer miners really begin to exploit the Alaskan deposits. In rapid succession, miners stampeded in 1898 to rich discoveries in the Nome area of Alaska, then in 1902 to the Fairbanks area; the Fairbanks placers were among the last of importance to be discovered.

Any history of placer mining would be incomplete without a word on dredging, which marked a major turn in operational efficiency. Dredging offered a way to handle tremendous quantities of material at a low unit cost and made it possible to mine where gold values were as little as a few cents per cubic yard

Probably the first successful bucketline dredge in the United States was operated in 1895 on Grasshopper Creek near Bannock in Beaverhead County, southwestern Montana. Others quickly followed, until by 1910, use of dredges had grown so that in California alone about 100 were in existence, of which 63 were reported in operation.

The first gold dredging in Alaska occurred about 1903, and the number of Alaskan dredges grew, until in 1914, 42 were in operation. The peak number of active dredges, 49, was not reached until 1940; World War II then interrupted most operations. Costs rose beyond profitable levels after the war, and only a few of the deactivated dredges were returned to service.

All gold dredges of any significance in the United States have been shut down, and most have been dismantled or sold abroad. Placer gold production today is primarily a byproduct of washing sand and gravel for use as an aggregate in the construction industry. Commercial placer mining by other means continues only at a few locations.

Total placer gold production in the United States from 1792 through 1968 is given in table 1. California and Alaska have accounted for more than three fourths of the total production of record. A large share of the overall production, it should be added, has come from dredges. The following list includes essentially all States and counties in which placer gold is known to occur (12):[2]

_____

[2]Underlined numbers in parentheses refer to items in the bibliography at the end of this report.

Alabama: Chilton, Clay, Cleburne, Coosa, Randolph, Talladega.

Alaska: For areas of occurrence, see figure 3.

Arizona: Cochise, Mohave, Pima, Pinal, Yavapai, Yuma.

California: Amador, Butte, Calaveras, Del Norte, El Dorado, Fresno, Humboldt,
    Imperial, Kern, Los Angeles, Madera, Mariposa, Mono, Monterey, Nevada,
    Placer, Plumas, Sacramento, San Luis Obispo, Shasta, Sierra, Siskiyou,
    Trinity, Tuolumne, Yuba.

Colorado: Adams, Boulder, Chaffee, Clear Creek, Costilla, Eagle, Gilpin,
    Hinsdale, Jefferson, Lake, Mineral, Moffat, Montezuma, Park, Routt,
    San Juan, San Miguel, Summit.

Georgia: Barrow, Bibb, Carroll, Cherokee, Dawson, Douglas, Fannin, Forsyth,
    Fulton, Gilmer, Greene, Haralson, Hart, Henry, Lincoln, Lumpkin, Madison,
    Marion, McDuffie, Murray, Newton, Oglethorpe, Paulding, Rabun, Towns,
    Union, Walton, Warren, White, Wilkes.

Idaho: Ada, Adams, Bannock, Benewah, Boise, Bonneville, Camas, Cassia, Clear-
    water, Custer, Elmore, Idaho, Latah, Lemhi, Owyhee, Power, Shoshone, Twin
    Falls, Valley, Washington.

Montana: Beaverhead, Broadwater, Deer Lodge, Fergus, Granite, Jefferson,
    Judith Basin, Lewis and Clark, Lincoln, Madison, Meagher, Mineral,
    Missoula, Park, Powell, Silver Bow.

Nevada: Clark, Douglas, Elko, Esmeralda, Eureka, Humboldt, Lander, Mineral,
    Nye, Ormsby, Pershing, Washoe, White Pine.

New Mexico: Colfax, Grant, Lincoln, Otero, Rio Arriba, Sandoval, Santa Fe,
    Sierra, Taos.

North Carolina: Anson, Burke, Cabarrus, Caldwell, Catawba, Chatham, Cherokee,
    Clay, Cleveland, Davidson, Franklin, Gaston, Granville, Guilford, Halifax,
    Henderson, Iredell, Lincoln, Macon, McDowell, Mecklenberg, Montgomery,
    Moore, Nash, Orange, Person, Polk, Randolph, Richmond, Rowan, Rutherford,
    Stanly, Union, Warren, Yadkin.

Oregon: Baker, Coos, Curry, Douglas, Grant, Jackson, Josephine, Union,
    Wheeler.

South Carolina: Cherokee, Chester, Chesterfield, Kershaw, Lancaster, Spartan-
    burg, Union, York.

South Dakota: Custer, Lawrence, Pennington.

Utah: Beaver, Daggett, Garfield, Grand, Piute, Salt Lake, San Juan, Sevier,
    Uintah.

Virginia: Albemarle, Buckingham, Culpeper, Cumberland, Fluvanna, Goochland,
    Louisa, Spotsylvania, Stafford.

Washington: Chelan, Clallam, Ferry, Kittitas, Lincoln, Okanogan, Whatcom.

TABLE 1. - Placer gold production, by States, 1792-1969

State	Placer gold production, thousand troy ounces	Rank as placer producer	Placer share of all gold produced in State, percent	Placer share of all gold produced in country,[1] percent
Alabama...............	15	18	30.0	Negligible
Alaska...............	21,130	2	70.2	6.7
Arizona.............	500	10	3.7	.2
California.........	68,470	1	64.3	21.7
Colorado..........	1,798	7	4.4	.6
Georgia...........	600	8	68.8	.2
Idaho.............	5,625	4	67.7	1.8
Montana...........	9,001	3	50.8	2.9
Nevada............	1,901	6	7.1	.6
New Mexico........	505	9	22.0	.2
North Carolina.....	245	13	20.5	.1
Oregon.............	3,500	5	60.2	1.1
South Carolina.....	52	15	16.3	Negligible
South Dakota.......	351	11	1.1	.1
Utah..............	75	14	.4	Negligible
Virginia..........	50	16	29.8	Negligible
Washington........	275	12	7.0	.1
Wyoming...........	43	17	97.6	Negligible
Total[2]........	114,136	-	-	([1])

[1] Based on total gold production of 316.77 million ounces, including 4.66 million ounces from undesignated sources. Placer share of all gold produced in country was approximately 37 percent.

[2] Exclusive of small production not identifiable by States.

Source: Based on data in table 6 of Bureau of Mines IC 8331 (24), with addition of production for 1965-69 as reported in the Minerals Yearbooks for those years.

WHERE TO LOOK FOR PLACERS

Figures 1 and 2 show general areas of the conterminous United States and Alaska, respectively, where placer gold has been produced. Placers can be found in virtually any area where gold occurs in hard rock (lode) deposits. The gold is released by weathering and stream or glacier action, carried by gravity and hydraulic action to some favorable point of deposition, and concentrated in the process. Usually the gold does not travel very far from the source, so knowledge of the location of the lode deposits is useful. Gold also can be associated with copper and may form placers in the vicinity of copper deposits, although this occurs less frequently.

Geological events such as uplift and subsidence may cause prolonged and repeated cycles of erosion and concentration, and where these processes have taken place, deposits may be enriched. Ancient river channels (referred to as

6

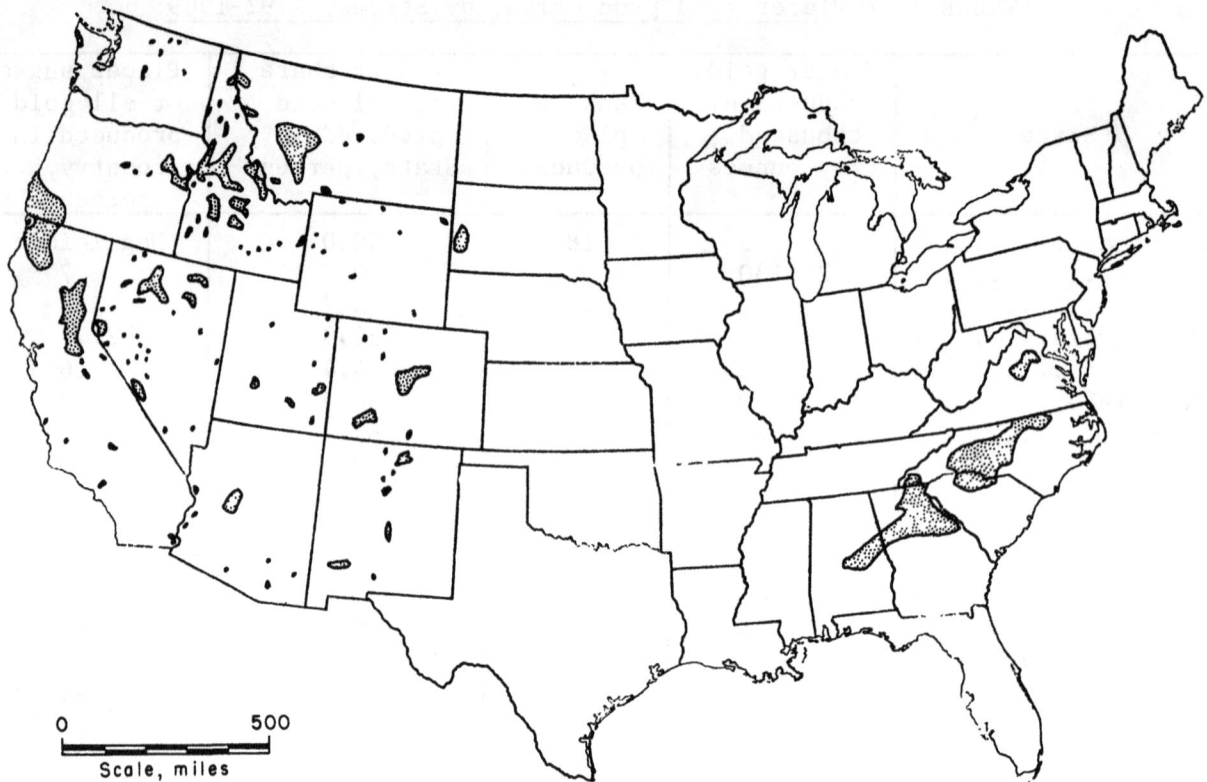

FIGURE 1. - Regions of the Conterminous United States Where Gold Has Been Produced From Placers. Note: See p. 4 for list of producing counties. Source: Reference 12.

the "Tertiary channels" in California) and certain river bench deposits are examples of gold-bearing gravels that have been subjected to a number of such events, followed by at least partial concealment by other deposits, including volcanic materials.

Residual placer deposits formed in the immediate vicinity of source rocks are usually not the most productive, although exceptions occur where veins supplying the gold were unusually rich. Reworking of gold-bearing materials by stream action leads to the concentrations necessary for exploitation. In desert areas deposits may result from sudden flooding and outwash of intermittent streams.

As material gradually washes off the slopes and into streams, it becomes sorted or stratified, and gold concentrates in so-called pay streaks with other heavy minerals, among which magnetite (black, heavy, and magnetic) is almost invariably present. The gold may not be entirely liberated from the original rock but may still have the white-to-gray vein quartz or other rock material attached to or enclosing it. As gold moves downstream, it is gradually freed from the accompanying rock and flattened by the incessant pounding of gravel. Eventually it will become flakes and tiny particles as the flattened pieces break up.

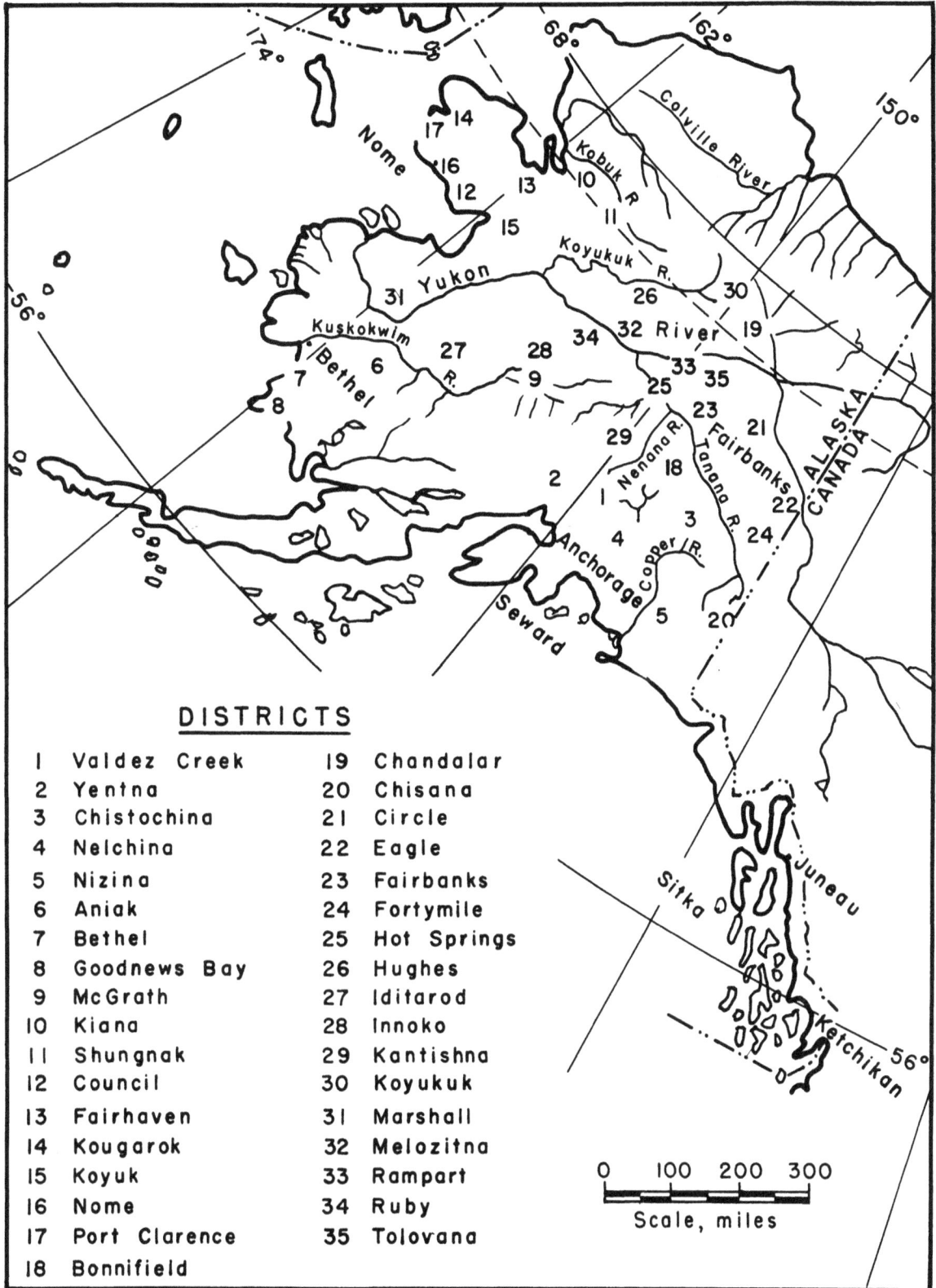

DISTRICTS

1	Valdez Creek	19	Chandalar
2	Yentna	20	Chisana
3	Chistochina	21	Circle
4	Nelchina	22	Eagle
5	Nizina	23	Fairbanks
6	Aniak	24	Fortymile
7	Bethel	25	Hot Springs
8	Goodnews Bay	26	Hughes
9	McGrath	27	Iditarod
10	Kiana	28	Innoko
11	Shungnak	29	Kantishna
12	Council	30	Koyukuk
13	Fairhaven	31	Marshall
14	Kougarok	32	Melozitna
15	Koyuk	33	Rampart
16	Nome	34	Ruby
17	Port Clarence	35	Tolovana
18	Bonnifield		

FIGURE 2. - Placer Mining Districts of Alaska. Source: Reference 22.

Some gold is not readily distinguishable by the normal qualities of orange-yellow to light yellow metallic color and high malleability, where it occurs in a combined form with another element, such as tellurium. Upon weathering, such gold may be coated with a crust, such as iron oxide, and have a rusty appearance. This "rusty gold," which resists amalgamation with mercury, may be overlooked or lost by careless handling in placer operations.

As mentioned before, the richest placers are not necessarily those occurring close to the source. Much depends on how the placer materials were reworked by natural forces. Streambed placers are the most important kind of deposit for the small-scale operator, but the gravel terraces and benches above the streams and the ancient river channels (often concealed by later deposits) are potential sources of gold. Other types of placers include those in outwash areas of streams where they enter other streams or lakes, those at the foot of mountainous areas or in regions where streams enter into broader valleys, or those along the ocean front where beach deposits may form by the sorting action of waves and tidal currents. In desert areas, placers may be present along arroyos or gulches, or in outwash fans or cones below narrow canyons.

Because gold is relatively heavy, it tends to be found close to bedrock, unless intercepted by layers of clay or compacted silts, and it often works its way into cracks in the bedrock itself. Where the surface of the bedrock is highly irregular, the distribution of gold will be spotty, but a natural rifflelike surface favors accumulation. Gold will collect at the head or foot of a stream bar or on curves of streams where the current is slowed or where the stream gradient is reduced. Pockets behind boulders or other obstructions and even moss-covered sections of banks can be places of deposition. Best results usually come from materials taken just above bedrock. The black sands that accumulate with gold are an excellent indicator of where to look.

It should be kept in mind that each year a certain amount of gold is washed down and redeposited during the spring runoffs, so it can be productive to rework some deposits periodically. This applies chiefly to the near-surface materials such as those deposited on the stream bars or in sharp depressions in the channels. The upstream ends of stream bars are particularly good places for such deposits. Where high water has washed across the surface by the shortest route, as across the inside of a bend, enrichment often occurs. A rifflelike surface here will enhance the possibility of gold concentration.

In prospecting areas with a history of mining, try to find places where mechanized mining had to stop because of an inability to follow and mine erratic portions of rich pay streaks without great dilution from nonpaying material. Smaller scale selective mining may still be practical here if a miner is diligent.

Placer gold occurs in so many areas that it would be impractical to try to identify each of them here. One of the best recent publications covering individual districts and areas is U.S. Geological Survey Professional Paper 610, Principal Gold Producing Districts of the United States, published in 1968 (14). Also, a series of reports is being written describing the individual placer gold deposits of various States or portions of States, to be published as Geological Survey Bulletins. For general areas of occurrence the maps in figures 1 and 2 may be consulted. Specific locations and names of mines can often be found on the detailed maps prepared by the U.S. Forest Service, the U.S. Bureau of Land Management, the U.S. Army Corps of Engineers, or the U.S. Geological Survey. Various State agencies may also have appropriate maps on hand.

California

California has led all other States in placer mining and as would be expected has many gold-producing areas of interest, particularly including the deposits on the Feather, Mokelumne, American, Consumnes, Calaveras, and Yuba Rivers and their tributaries (2, 4-6, 9). These rivers reach into the famous Mother Lode area of the Sierra Nevada Mountains from which much of the gold is derived. Deposits are also found in the drainages of the Trinity and Klamath Rivers in northern California and at scattered points in the southern part of the State. Ancient Tertiary channels and gravels of the Sierra Nevada Range have been especially productive sources of gold, and maps have been published by the California Division of Mines and Geology showing approximate routes of these features. Two U.S. Forest Service maps that the prospector would find of particular value in considering the Sierra deposits are of (a) the Downieville, Camptonville, and Nevada City districts, Tahoe National Forest; and (b) the Foresthill and Big Bend districts, also in the Tahoe National Forest. Maps covering the Trinity and Klamath National Forests of northern California might also be of interest.

Alaska

Most of Alaska's gold production has come from placers, principally those in the Yukon River Basin, although deposits are known on nearly all major rivers or their tributaries (11, 19, 22, 33). Beach deposits in the Nome district have been notably productive, as have the river and terrace or bench placers in the drainages of the Copper and Kuskokwim Rivers. Figure 2 shows the main placer districts of the State. Climatic conditions play a great part in mining in Alaska, and the season for hydraulic operations of any kind is relatively short.

Northwest States (Montana, Idaho, Oregon, and Washington)

Montana's principal placer mining districts are in the southwestern part of the State (15). The Helena mining district and the many placers along the Missouri River in the vicinity of Helena and upstream are among the more important areas. The headwaters and tributaries of the Missouri in Madison County, particularly near Virginia City and Bannock, are noted for early placer production. Placer gold has also been produced on the headwaters of the Clark Fork of the Columbia River at a number of points.

The Boise basin, northeast of Boise, Idaho, is most noted for the dredging of placers (21). Other well-known placer areas lie along the Salmon River in Lemhi and Idaho Counties and on the Clearwater River and its tributaries, particularly in the vicinities of Elk City, Pierce, and Orofino. Placer gold is also found along the Snake River, but this is commonly fine-grained or "flour" gold that is difficult to recover.

Oregon's placers are located mainly in the southwestern part of the State, on tributaries of the Rogue River and on streams in the Klamath Mountains (16). Main gold-producing areas are the Greenback district in Josephine County and the Applegate district in Jackson County. Placer gold also occurs in many of the streams that drain the Blue and the Wallowa Mountains in northeast Oregon. The Sumpter area and the upper Powder River have had important production. Other areas include the Burnt River and its tributaries and the John Day River Valley.

Washington is not noted for placers, although gold has been found along a number of its streams, including some on the western slope of the Cascade Mountains. Generally, the few productive placers have been confined to the north-central part of the State.

Nevada

Nevada has not been a large placer gold producer, although lode gold deposits--potential sources--are widely distributed throughout the State (27). The problem has been chiefly one of too little water. In the past, dry washers were used extensively, as well as other methods that were very conservative of water. Producing areas were largely found in the western half of the State and included American Canyon and Spring Valley in the Humboldt Range, Pershing County, and the Manhattan and Round Mountain areas of Nye County. Placers were also worked below Virginia City and in northern Elko County near Charleston. Signs of limited placer diggings may be seen in many parts of the State.

Colorado

A few important Colorado placers of the residual type are found on slopes and hillsides in the immediate vicinity of gold veins. However, placers in Colorado are generally confined to narrow canyons below lode gold mining areas within the Rocky Mountains in a belt which extends northeast across the western part of the State (28). Almost every gold district has had some placer production. Many of the streams emerging from the Front Range, the headwaters of the South Platte River, and the Arkansas River and its tributaries as far upstream as California Gulch contain placer gold. Historically, placers were mined first and led to development of Colorado's rich lode deposits.

Other States

Among the other Western States, placer mining has been limited to only a few localized areas. In South Dakota, the Black Hills (particularly the Deadwood area) and French Creek, near Custer, have been productive sources. Arizona (32) and New Mexico (30) placers are in some instances related to copper deposits that carry gold.

In the Eastern States some of the streams draining the eastern slopes of the southern Appalachian Range have yielded gold (1, 17). Saprolite deposits (rock decomposed at the original site) have been a source of placer gold in Georgia (13) and the Carolinas (3). Generally, the eastern placers are sparsely distributed and the gold is low in grade. Thus, few serious efforts have been made at mining them since the early 1800's. Nevertheless, many locations offer possibilities for small operations intended primarily for recreational purposes.

OTHER THINGS YOU NEED TO KNOW

Staking Claims on Open Lands

Instructions on staking claims and filing for patents can be obtained from the U.S. Bureau of Land Management. In addition to Federal regulations, individual States also have certain requirements pertaining to the location of claims on public lands. Information on these requirements is available from the State agency that deals with mining. Claims for mining can only be staked on lands of the public domain.

Lands in areas generally subject to location for mining, such as National Forests, may be open or closed depending on whether the land has been withdrawn for some special purpose. The status of the land can usually be determined by local inquiry to the U.S. Forest Service, or it can be checked at the Land Offices of the Bureau of Land Management or at the County Assessor's Office. The status of land being considered for mining should be established before any significant investment of money or labor has been made. This will insure that the ground is open to location so that the prospector can stake a valid claim and protect his investment.

When entering any land to examine for or attempt to mine placer gold, a person should determine if it is privately owned, previously located by claim that may still be valid, or possibly held under patent, which conveys the right of private ownership. Under any of these conditions, the unauthorized intruder is trespassing and has no legal rights to the gold he may produce. Usually some sign or indication of ownership is evident, or a local resident can supply the information necessary to determine ownership, but in any event one takes his chances when the status of land is uncertain. At the worst, the land may be protected by a shotgun. Active claims should be clearly marked, and records may be checked in the respective County Courthouse to determine approximately where and when they were located. It should be added that, however necessary, such checking can be tedious.

The basic laws on location of mining claims in the public domain are contained in the General Mining Laws of 1872. Placer claims generally can be located on lands that would be classified as locatable if they contained vein or lode deposits. Neither the beds of navigable lakes and rivers nor lands below high tide are subject to mineral location. However, new claims can be located over abandoned earlier ones, although the new locator may be called upon to establish that the earlier claims were, in fact, abandoned at the time of relocation.

Mining locations may be made by U.S. citizens, by those who have declared their intention to become citizens, by an association of qualified persons, or by a domestic corporation. A location may be made by a minor who has reached "the age of discretion," and without regard to the sex or residence of the locator. A person may make valid locations as an agent for other qualified parties who may not have even seen the ground. No limit is placed by Federal statutes on the number of locations that may be made, and claims may be amended and boundaries changed at any time, provided such changes do not interfere with the rights of others.

Generally, a placer claim is established by posting a notice of location upon a tree, a rock in place, a stone, a post, or a monument. This must contain the name of the claim, name of locator or locators, date of location, number of square feet or acreage claimed, and sufficient description of the claim by reference to some natural object or permanent monument to identify the claim, following which the boundaries of the claim must be marked so they may be readily traced. Requirements for marking of boundaries vary somewhat by State. A location may not exceed 20 acres for any one person nor 160 acres for an association of persons, and claims should conform as nearly as practicable with the rectangular subdivision characteristic of the U.S. system of public land surveys.

At least one discovery of mineralization is required per claim (20 to 160 acres) "that would justify a person of ordinary prudence in the further expenditure of time and money, with reasonable prospect of success in developing a profitable mine."[3] A discovery implies a certain amount of excavation to show that the required mineral is indeed present, although State laws vary somewhat on this point. A minimum annual expenditure in labor and improvements of $100 must be made to hold possession of a mining claim, and evidence to this effect must be duly recorded by the appropriate county recorder. Such work is generally known as "assessment work." Provisions are also made for millsites on nonmineral land. Since September 2, 1958, the requirements for assessment work may be satisfied by conducting geological, geochemical, and geophysical surveys under the supervision of a qualified expert.

Unpatented claims may be bought and sold, but their use is restricted to mining purposes only. Any use of the surface for purposes unrelated or foreign to mining is unauthorized. Ownership of both the minerals and lands prospected and developed is attained by the process of patenting. Prior to patenting, the claim holder has possessory rights only to the minerals.

A number of requirements must be met for a patent, which is obtained through the U.S. Bureau of Land Management. These include proof of discovery of valuable minerals, expenditure of at least $500 in labor and improvements, payment for filing application, publishing costs, and other items. Specific information on placer mining patent applications and on adverse claim procedures in the case of contested rights can be found under parts 3863 and 3870,

---

[3]For assistance regarding location of mining claims on the public domain, the Bureau of Land Management of the U.S. Department of the Interior should be consulted.

title 43, of the Code of Federal Regulations pertaining to mining claims under the General Mining Laws of 1872.

Public Law 167, enacted July 23, 1955, provides for multiple use of the surface of public lands. This does not alter the validity of gold placer locations based upon sufficient evidence of discovery. It does set up procedures whereby the Government agency responsible for administering surface resources can challenge a mining claimant. In this way the question of surface rights is cleared so that the fullest use can be made of the land.

Under Public Law 359, enacted August 11, 1955, mining is permitted on lands which have been withdrawn from location or reserved for power development and for other purposes, provided certain steps are taken. In this case, permission to mine must be obtained from the Secretary of the Interior. Claims located before the date of the act on a power withdrawal are relocatable.

Lands accorded to Indian Reservations are not subject to claim under U.S. mining laws. (From June 18, 1943, to May 27, 1955, an exception was made for the Papago Indian Reservation, but such free mineral entries were then curtailed (43 CFR 3825, formerly 3635).) One wishing to prospect or mine on Indian lands must obtain a lease from the Secretary of the Interior; application for the lease should be made through the reservation superintendent.

On State-owned lands, application for prospecting or mining lease should be made to the appropriate State authority. Regulations on granting such leases vary by State. Mining claims can be located on private stock-raising homestead lands where the Government has reserved rights to minerals, although certain limitations are specified to protect the homesteader and to provide reimbursement for damage to crops or to tangible improvements. Taylor Grazing Act lands are also subject to location.

Privately owned lands are usually leased or may be purchased outright for mining purposes. Normally a lease will carry a royalty provision on the mineral production as a percentage of the gross or net value received from sales. Many different arrangements are possible, depending upon the requirements or bargaining positions of the potential lessor and lessee.

The recent completion of studies of mining laws by the Public Land Law Review Commission and recommendations by the Commission can be expected to result in changes in regulations that could affect placer location procedures or mining rights. Information on such changes as they occur will become available from the Bureau of Land Management, U.S. Department of the Interior.

Problems With Water Rights, Water Supply, and Stream Pollution

The need for a good, dependable, and plentiful supply of water increases geometrically with the scale of operation in placer mining. Panning gold requires very little water and can be done in a small tub if necessary. At the other extreme, the hydraulic monitor, once in use, employed large flows of water under high pressure, and sluicing at a large operation could consume

virtually all the water that might be available. One thing the placer miner
must keep in mind is the seasonal nature of stream flow. This affects both
the supply of water and also the problems of pollution for downstream users
and damage to stream ecology.

Various means are used to divert and impound water. Channels, pipes, and
flumes can be constructed to conduct water where it is wanted. If supply at a
continuous flow is limited, storage must be provided, and placer operation is
then restricted to periodic activity and depends on the capacity of the reservoir. A simple tank may make a suitable reservoir for a small operation.
Pumps are commonly used now where power is cheap enough, and the recirculation
enables use of a smaller supply of water.

The question of water rights has always been important to the placer
miner and is a complex subject in itself. Legal authorities should be consulted in case of any doubts or disputes. It has been common practice in
placer mining to measure water requirements in terms of "miner's inches,"
which can be converted to rate of flow by the approximate factor of 40 inches
to 1 cubic foot per second (the legal conversion in Arizona, California, Montana, and Oregon; other States vary). One cubic foot of water is equivalent
to about 7.5 gallons. Thus, a miner's inch converts to 11.2 gallons per minute.

Water flows are measured in cubic feet per second and storage is measured
in acre-feet, the latter being equivalent to a 1-foot depth of water spread
across an area of 1 acre. Measurement of flow is usually done with a cali-
brated weir, but flows can be estimated by average velocity or by other methods
if the quantity is large. A term used to describe the effectiveness of water
in hydraulic operations is its "duty," which is usually expressed as the number of cubic yards of gravel washed per miner's inch per 24-hour day. Water
duty will vary greatly with the mining situation.

States where placer mining has been important in the past, such as California, have enacted detailed and quite strict laws regarding stream pollution
from placer operations. Such laws require the construction of a settling pond
or ponds of sufficient size to clarify the water used in mining before it is
discharged into the stream. Furthermore, they may require that aluminum sul-
fate and lime or some similar clarifying substance be added to the effluent to
avoid rendering the water in the stream unfit for domestic water supply
purposes.

Regulations regarding stream pollution may vary with the stream and the
particular portion of the State, so the appropriate control agencies should be
consulted. In California, the California Debris Commission and the California
Department of Fish and Game regulate discharges from mining operations. The
California Fish and Game Code curtails mining operations in the Trinity and
Klamath River fish and game district between July 1 and November 30, "except
when the debris, substances, tailings, or other effluent from such operations
do not and cannot pass into waters in the said district."[4] Federal and State

---

[4] Quotes in the following sections are obtained from relevant sections of
   California statutes.

water resource and water quality control agencies may also have something to say about placer mining discharges, and it is wise to check with them before undertaking any sizable project.

Legislation in California that closed down hydraulic mining on the Sacramento River and its tributaries goes back many years. An act entitled "Protection of Domestic Water Supplies From Pollution of Placer Mining Operations" covering the watersheds of the Sacramento and San Joaquin Rivers (A. B. 2006) was passed in 1941, requiring a permit for placer mining from the California Debris Commission and compliance with a number of provisions specifically aimed at dredging.

Who Can Advise You

There are many sources to which the novice placer miner may turn for information. Probably the first should be the particular State agency dealing with geology, mining, conservation, or development. Universities with geology or mining departments will have knowledgeable people who can be consulted. A readily accessible source of information is the reference section of your public library. Professional engineers and consultants may be contacted through professional organizations or directly, by telephone.

Federal agencies most concerned with the problems of placer mining are the Bureau of Mines, the Geological Survey, and the Bureau of Land Management, Department of the Interior. Generally, the Bureau of Mines is best equipped to handle technical or statistical questions; the Geological Survey provides information on geology and deposits; and the Bureau of Land Management is involved in land ownership and evaluation problems. The Forest Service, Department of Agriculture, is a good source of information on placer mining sites and regulations within the National Forests. Furthermore, the prospector is well advised to inquire at the local ranger station of the National Forest in which he intends to prospect for guidance and to inform the ranger of his intent.

County offices, including that of the County Recorder, can often supply useful information on claim staking and placer locations within that county. Forms for location notices are normally available at the County Courthouse in mining localities or may be purchased at a stationery store. Questions about possession and sale of gold come under the purview of the Department of the Treasury.

HOW TO LOOK FOR PLACERS

Once decided on the area of search and armed with some knowledge of the characteristics of deposits to look for, then you are ready to explore. Most areas are relatively settled today and are accessible by car, or at least lie within a few miles of a road. A possible exception is Alaska, where an aircraft or boat might be needed to reach the site. Regardless of the type of transportation, you will need adequate supplies and equipment to sustain you and your companions for an extended stay in the field. With a gold pan for each and setting out from a base camp it should be possible to determine within several days if the potential for the area is good.

Equipping Yourself

Camping and outdoor recreation in general have become so popular that many commercial sources of equipment and information are now available. Some stores appeal to the budget-minded, while others, such as the specialty shops for camping supplies, have a wider selection of usually more durable products. Books on camping are available at the library, and reliable merchants will recommend the equipment best suited for a particular use. Many of the comforts of home can be found in the ordinary camp today. Backpacking has benefited from developments in lightweight materials and foods. The amounts and types of goods and equipment selected will depend on the remoteness of your location and accessibility of a resupply point. The prospector might wish to travel with a mobile home, trailer, or camper, or he might simply pack his gear on his back and head up the trail. A few suggestions are in order here, but the individual must do much of his own planning, since requirements and tastes vary so greatly.

Basic Equipment

Among the essential implements needed for prospecting are a pick; a long-handled, round-pointed shovel; and a gold pan, preferably a 10- or 12-inch-diameter pan which can usually be purchased at hardware stores in gold-mining areas. A small prospector's pick is also useful, and a magnet and a small amount of mercury should be carried to separate the gold from black sand after panning. Specialty stores and manufacturers can provide the more elaborate equipment, such as skindiving gear, ready-built sluices, and mechanical gold separation devices, if desired.

In some cases, a bucket or wheelbarrow may be needed to transport materials to the washing site, and in addition, a heavy 1/4- to 1/2-inch-mesh screen is handy to separate out coarse materials. A small screen cut to nest in the upper part of a gold pan can be useful for the same purpose in panning. A gold pan the same size as the one used for panning will make a most efficient nesting screen if a close pattern of holes is drilled in the bottom. Holes usually should be 1/4 to 3/8 inch in diameter, depending on the average size of the material being sampled. Distance between holes should be about the same as the diameter of the holes. In some areas these pans can be purchased readymade. For weighing gold, a small balance scale graduated in milligrams may be desirable. A compact, folding type of balance is available for this purpose.

A compass will be needed for establishing claim lines and for finding your way out of the woods if lost. Adequate maps should be carried. A hand magnifying lens is helpful in identifying minerals. Bags may be needed to carry out samples; plastic bags are the best because samples may be damp. A rocker may be transported to the site either assembled or in a knocked-down condition. If mining is planned, lumber and other materials to build a sluice may be carried to the site. (See construction details under respective headings.) More elaborate equipment such as pumps, pipes, hoses, and light plants might be taken in by pack animals if desired.

Personal gear includes a <u>good</u> pair of boots, sturdy clothing, weather-proof gear, sleeping bag, tent, and such other things as one might want for comfort and sanitation. A foam pad or air mattress adds comfort to sleeping. A length of rope is useful for many purposes around camp, from raising the food out of reach of animals to extracting a car from a mudhole. For hiking, all necessary equipment for the period away from camp should fit into a manageable backpack of some kind.

An ax, a flashlight, a knife, and matches are almost indispensable. (Fires in the National Forest should be made only in designated areas or after consulting the local forest ranger.) A water bucket is often required, and a good crosscut saw will be found useful. Guns and fishing equipment can be taken to supplement the food supply and to provide some additional recreation. Guns are seldom necessary for protection from animals. A canteen with a 2-quart or larger capacity is advisable in many areas, depending on dryness of the climate. You will need water-purification tablets where streams are con-taminated, whether by grazing stock or for other reasons. A miner's lamp, which consumes calcium carbide, is sold at some hardware stores and can be used for a serviceable light, although most people when away from electricity prefer gasoline or propane lamps. A carbide lamp will also be useful for any underground work. The special miner's safety lamp is recommended wherever air may be bad. Stoves that burn gasoline or pressurized gas are in wide use in camping and even gas refrigerators may be taken along "to cool the beer." (For low-budget operations, a swift-running stream will serve this same pur-pose well.) For any length of time in the field, an oven for baking is a val-uable amenity. A reflector oven for use next to a campfire can be made of light sheet metal and will give excellent results, also serving as a place to keep food warm.

Supplies

Freeze-dried foods are generally good and easy to carry and prepare, although somewhat more expensive than most other foods. For estimating pack weights, about 2 pounds of dehydrated and freeze-dried foods is needed per person per day. Canned foods should be avoided when backpacking because of their weight, but they are otherwise satisfactory. Disposal of empty con-tainers should be done with consideration to others who may follow and wish an uncluttered landscape; burial is usually recommended.

Suggested food supplies for a prospector's camp include the following: bacon, beans, cheese, salt, baking powder and soda, coffee, tea, onions, potatoes, fruits, corn, peas, raisins, rice, flour, crackers, cereals, butter or margarine, powdered milk, eggs, pancake and waffle mix, sugar, syrup, and fresh meat and vegetables as practicable. Many other items can be added to the list, but these are most of the basics. Utensils should include a variety of dishes, silverware, a sharp knife, spatula, can opener, frying pan, coffee pot, and several different sizes of pots and pans. Towels, both paper and cloth, soap, scouring pads, and metal or plastic tubs or basins will be needed for cleaning up.

Extra clothing should be included in your supplies for warmth and for changes. Mosquito netting may be a virtual necessity in some areas, and ade-quate amounts of a good insect repellent should be packed.

Safety Needs

Probably the most troublesome and at times the greatest hazard in the wilds today is the bear. People may argue which type of bear has the meanest temperament, but any type may leave your camp a shambles when in search of food, and under certain circumstances any bear will attack a person. Placing food out of reach or in a secure container will help reduce the attraction. Fortunately, most bears will turn and run when frightened by loud noises. Other wild animals are seldom dangerous except when provoked, but smaller ones such as packrats can inflict considerable damage on camp gear and foodstuffs. Poisonous snakes, spiders, ticks, scorpions, and the like should be treated with traditional caution; their presence should be anticipated in most areas. Learn to identify and avoid poison oak and poison ivy! Knowledge of first aid is essential for dealing with emergencies that might arise on an outing, and a study or review of the subject should be included in any preparations.

Some of the personal hazards faced in the out-of-doors include twisted ankles, lacerations from falling in brush, falls from slippery rocks or crude bridges when crossing streams, breaking through floors in old building ruins, and falls or cave-ins in old mine workings. Beware of bad air in any old workings! Danger of drowning is always present when working around the deeper streams or pools when placer mining.

Many types of first aid kits and equipment are on the market. The choice of kit is one of size and variety of content. A snakebite kit is usually a separate accessory and should be carried, even though it is rarely put to use. Disinfectants, aspirin, fungicides, bandages, and similar items should be included. For areas of considerable sunshine, tanning lotion, sunglasses, and a hat are needed, and salt tablets should be taken as designated to prevent heat prostration. Wearing a safety hardhat and safety glasses may be advisable at times.

Panning for Gold

The standard gold pan is made of stiff sheet iron and is 16 inches in diameter at the top and 2-1/2 inches deep. The rim is flared outward at an angle of about 50° from the vertical. Smaller pans are used for testing, and it is advisable for most panners to use either a 10- or 12-inch size for handling ease. Probably the 12-inch is the most widely available. Frying pans or other cooking utensils may also be used for washing out gold but are less effective. Before any kind of container is used for panning it should be cleaned thoroughly and all grease should be burned out. New pans generally are greasy and should be heated over a fire until this coating is gone. Even a rusty pan, if clean, can be used satisfactorily. In fact, the roughness due to the pitting of the rust may assist in holding back the gold.

There are different techniques and subtle variations in the art of panning--experience teaches which is best. Those with wide experience and much practice can recover the most gold with the least effort. It is sometimes said that good panning technique lies in the action of the wrists. After much practice the good panner should be able to save even the very fine gold that may be nearly but not quite free from the black sands.

The pan usually is filled level with the top, or slightly rounded, depending somewhat upon the nature of the material being washed and the personal preference of the panner. It is then submerged in water. Still water 6 inches to 1 foot deep is best. While under water the contents of the pan are kneaded with both hands until all clay is dispersed and the lumps of dirt are thoroughly broken. The stones and pebbles are picked out after the fines are washed off. Then the pan is held flat and shaken under water to permit the gold to settle to the bottom. The pan is then tilted and raised quickly--still under water--so that a swirling motion is imparted and some of the lighter top material is washed off. This operation is repeated, occasionally shaking the pan under water or with water in it until only the gold and heavy minerals are left. With proper manipulation, this material concentrates at the edge of the bottom of the pan. Care must be taken that none of the gold climbs to the lip of the pan or gets on top of the dirt.

Nuggets and coarse colors of gold can now be picked out readily with a tweezer or with the point of a knife. Cleaning the black sand from the finer gold is more difficult, but can be carried nearly or entirely to completion by careful swirling of the contents as described above, always watching to see that none of the colors are climbing toward the lip. This part of the operation usually is done over another pan or in a tub so that if any gold is lost it can be recovered by repanning.

The concentrates should be dried, and the black sands (composed largely of magnetite) can then be removed by a magnet or by gently blowing them on a smooth flat surface. If there is an excessive quantity of black sand, the gold usually is amalgamated by putting a portion of a teaspoonful of mercury in the pan. In sampling work, extra care should be taken to see that no fine colors are lost. When mining, however, additional time needed to insure that all colors are saved probably is not justified because the value they add is so small.

A word should be said here about other minerals that you may see in your gold pan. Pyrite ("fool's gold," an iron sulfide) and mica are often mistaken for gold by the novice. Pyrite, which is usually a brassy yellow to white color, will shatter when struck with a hammer and becomes a black powder when finely ground. Mica, which may have a bright, bronzy appearance, is distinguished by its light weight and flat, platy cleavage. Both minerals are common in gold areas. Other minerals that will collect with the gold and black sands because of high specific gravity include ilmenite (iron-titanium oxide), hematite (nonmagnetic iron oxide), marcasite (an iron sulfide), rutile (titanium oxide), scheelite (calcium tungstate), wolframite (iron, manganese tungstate), tourmaline (boron and aluminum silicate), zircon (zirconium silicate), chromite (iron and chromium oxides), and cinnabar (mercury sulfide). If present in sufficient quantity, these latter minerals may have some economic significance, although efforts to recover them as byproducts are seldom worthwhile. Native platinum, elemental mercury, lead shot, and similar materials are also occasionally found in the pan.

EVALUATION: SHOULD YOU INVEST AND MINE?

This question becomes more difficult to answer as the size of the planned operation increases. Estimation of the amount of gold recoverable and the overall costs of investment and mining is no simple matter and calls for highly experienced engineering skills for any moderate- to large-scale project. Elaborate procedures of sampling and evaluation cannot be followed by the small-scale operator because of the cost. Thus, his decisions must be based on a variety of factors, not the least of which is intuition. Needless to say, many mistakes have been made, with much resultant waste of money and effort. Do not let what started out as a recreational activity become your master instead of your servant.

Sampling Techniques

Many methods of sampling are possible, including the simple panning of gravel from surface exposures, churn drilling, test pitting and trenching, shaft sinking, and drifting. As an aid in tracing possible gold-bearing channels, geophysical techniques have been employed with some success, but proper use of the typical instruments involved is generally reserved to experts. Moreover, interpretation of results is seldom adequate to provide any quantitative estimates, although the information gained can be useful in planning an exploration program.

For a thoroughgoing discussion of exploration and sampling techniques the reader is referred to the recent Bureau of Land Management publication by John H. Wells, entitled Placer Examination: Principles and Practice (31). Wells' description of panning is particularly recommended.

Panning and rocking (described later) are the basic means of determining the recoverable gold content of placer materials. A fire assay, sometimes made on a concentrate, provides a relatively complete estimate of the gold content of the material, but a poor estimate of how much gold can actually be extracted by conventional washing methods. Thus, placer gold is seldom assayed, except to determine its fineness (measure of gold purity, see p. 37). In estimating the value of gold in the pan after washing a quantity of gravel, the technique of counting nuggets and "colors" is normally followed. Generally, pieces worth more than 5 or 10 cents are considered as nuggets; smaller particles are colors. When skill is developed in estimating the various sizes of particles, a good degree of consistency can be achieved in the results.

Where samples can be obtained across a section of the bank exposed along a creek, it is good practice to cut a vertical groove or channel of fairly consistent width and depth. The sample may be cut from top to bottom, or in segments comprising several different samples if the bank shows distinct changes in materials. Bars may be sampled by digging a vertical hole, clear to bedrock if possible, and panning the product. For surface mining of "skim bars," sampling consists of simply taking a panful from a favorable point and visually estimating the amount of similar material in the vicinity. Clearly, there is not much accuracy in any of these methods, but the deposition of gold in such locations is bound to be erratic anyway. More representative sampling is usually possible in the larger deposits where deposition and size of gold particles is more uniform or consistent.

Calculating What You Might Have

For the small-scale miner, sampling will usually be limited to taking a panful here and there and possibly running a larger sample through a rocker or sluice if panning discloses any gold. If colors are found, a record should be made of the number and estimated size of colors per pan and the approximate location. The sampling then progresses until one is assured the prospects are good enough to warrant a mining operation of some sort.

A scale of sizes and approximate values of colors based on pure gold at $35 an ounce is as follows: (Note: Mesh = screen size in openings per square inch; minus 10- plus 20-mesh material will pass screen with 10 openings per square inch but be stopped by screen with 20 openings.)

Coarse gold, plus 10 mesh: should be picked out and weighed individually, value about $1 per gram.

Medium gold, minus 10 mesh but plus 20 mesh: 2,200 colors per troy ounce, value about two-thirds of a color to 1 cent.

Fine gold, minus 20 mesh but plus 40 mesh: 12,000 colors per troy ounce, value about 3 colors to 1 cent.

"Flour" gold, minus 40 mesh: 40,000 colors per troy ounce, value about 10 colors to 1 cent.

Differing fineness or price will affect the values somewhat.

It is common to report panning results in cents per pan. So, assuming you have determined that a "pan factor" of about 400 pans per cubic yard (bank measure) for the 12-inch pan is a suitable figure, multiplying the cents-per-pan figure by 400 gives the estimated value per cubic yard.

Another means of estimating is to rank the colors into three groups, as follows:

Number 1: colors weighing over 4 milligrams

Number 2: colors weighing between 1 and 4 milligrams

Number 3: colors weighing less than 1 milligram

(Note: 31,103 milligrams equals 1 troy ounce.)

Scales will be needed to check the weights until the eye can judge the sizes properly. It is recommended that particles over 10 milligrams be weighed individually. A rough measure of value is one-tenth of a cent per milligram. Thus, the value in a pan can be calculated using your visual count and tally of the number of colors of each rank. After sufficient practice, good estimates will come easily. Thickness has a great bearing on weight: For instance, some gold might look large, but actually be flat, flaky, and hence very light.

Determining the overall value of a deposit with any accuracy calls for a knowledge of accepted practices and mathematical procedures for weighting the values and sample intervals. It is important also to understand the statistical principles of variation and distribution, which are beyond the scope of this report. Generally, the practical prospector will take a few measurements, make some crude calculations using his panning results, and decide to stay or move on.

HOW TO GO ABOUT MINING

When a site where gold is known to occur has been found, and after it has been sampled and judged worthy of further effort, the ownership status should be checked to assure that the ground is open for claiming. Then, after staking adequate claims (or arranging to lease if the ground is not open to claim), you are ready to consider mining. Whether mining permits are required by those State agencies involved with fish and game or watersheds should be investigated, because placer operations of any size may drastically change the local water quality. A simple operation may have virtually no effect on a stream or surroundings, but when materials amounting to more than a few cubic yards a day are handled, the possible effects begin to become significant.

Choosing a Method

Among the simpler hand methods of recovering gold are the gold pan, the rocker, the dip-box, the long tom, and the sluice. Panning has been described in a previous section, entitled "How to Look for Placers," and will only be discussed briefly here. The pan is generally too slow to be effective for anything more than prospecting. The rocker is a time-honored device of the small-scale miner with limited means. The dip-box and long tom might be considered more like simplified sluicing methods than distinct methods in themselves. As a method, the long tom has never been very popular but is described here for its possible historical interest. Other methods used in specific circumstances would include the surf washer, the dry washer, and skindiving.

The simpler methods all normally involve hand-mining operations (shoveling and/or picking of the gold-bearing materials). Limited mechanization is sometimes practical for moving and washing gravels in even the smallest operation, and this possibility should not be overlooked. Even motorized devices for panning are marketed by several manufacturers. Pumps and small excavators can often be adapted to the small mining operation by the enterprising miner.

The more complex methods, such as ground sluicing, hydraulicking, drift mining, excavation using powered equipment, and dredging, require considerable investment, knowledge, and experience; a full discussion of these methods is beyond the scope of this report.

The choice of method depends primarily on the scale of operation and the availability of water. These and other characteristics of the different methods are discussed below.

Gold Pan

Panning is the hardest way to wash gold from placer gravels, but it is an inexpensive and completely mobile method. A person can dig with a pick and shovel much faster than he can pan the material dug, so it pays to treat only the highest grade products by panning once one has settled down to mining.

An experienced person can wash about 10 large pans per hour, the equivalent of approximately 1/2 to 1 cubic yard of gravel per day, depending on how clean the gravel is. A level-full, standard 16-inch pan might contain roughly 22 pounds of dry bank gravel; there are approximately 150 to 180 pans per cubic yard of gravel. More than twice as many 12-inch pans would be required per cubic yard. The top dirt or cover is usually cast aside and the few inches of material directly above bedrock and the material scraped from crevices is panned. Places to look and the proper panning technique have been covered in earlier sections.

Rocker

At least twice as much gravel can be worked per day with the rocker as with the pan. The rocker or cradle, as it is sometimes called, must be manipulated carefully to prevent loss of fine gold. With the rocker, the manual labor of washing is less strenuous, but whether panning or rocking, the same method is used for excavating the gravel.

The rocker, like the pan, is used extensively in small-scale placer work, in sampling, and for washing sluice concentrates and material cleaned by hand from bedrock in other placer operations. One to three cubic yards, bank measure, can be dug and washed in a rocker per man-shift, depending upon the distance the gravel or water has to be carried, the character of the gravel, and the size of the rocker. Rockers are usually homemade and display a variety of designs. A favorite design consists essentially of a combination washing box and screen, a canvas or carpet apron under the screen, a short sluice with two or more riffles, and rockers under the sluice (fig. 3). The bottom of the washing box consists of sheet metal with holes about 1/2 inch in diameter punched in it, or a 1/2-inch-mesh screen can be used. Dimensions shown are satisfactory but variations are possible. The bottom of the rocker should be made of a single wide, smooth board, which will greatly facilitate cleanups. The materials for building a rocker cost only a few dollars, depending mainly upon the source of lumber.

After being dampened, the gravel is placed in the box, one or two shovelfuls at a time. Water is then poured on the gravel while the rocker is swayed back and forth. The water usually is dipped up in a simple long-handled dipper made by nailing a tin can to the end of a stick. A small stream from a pipe or hose may be used if available. The gravel is washed clean in the box, and the oversize material is inspected for nuggets, then dumped out. The undersize material goes over the apron, where most of the gold is caught. Care should be taken that not too much water is poured on at one time, as some of the gold may be flushed out. The riffles stop any gold that gets over the apron. In regular mining work, the rocker is cleaned up after every 2 to 3 hours, or

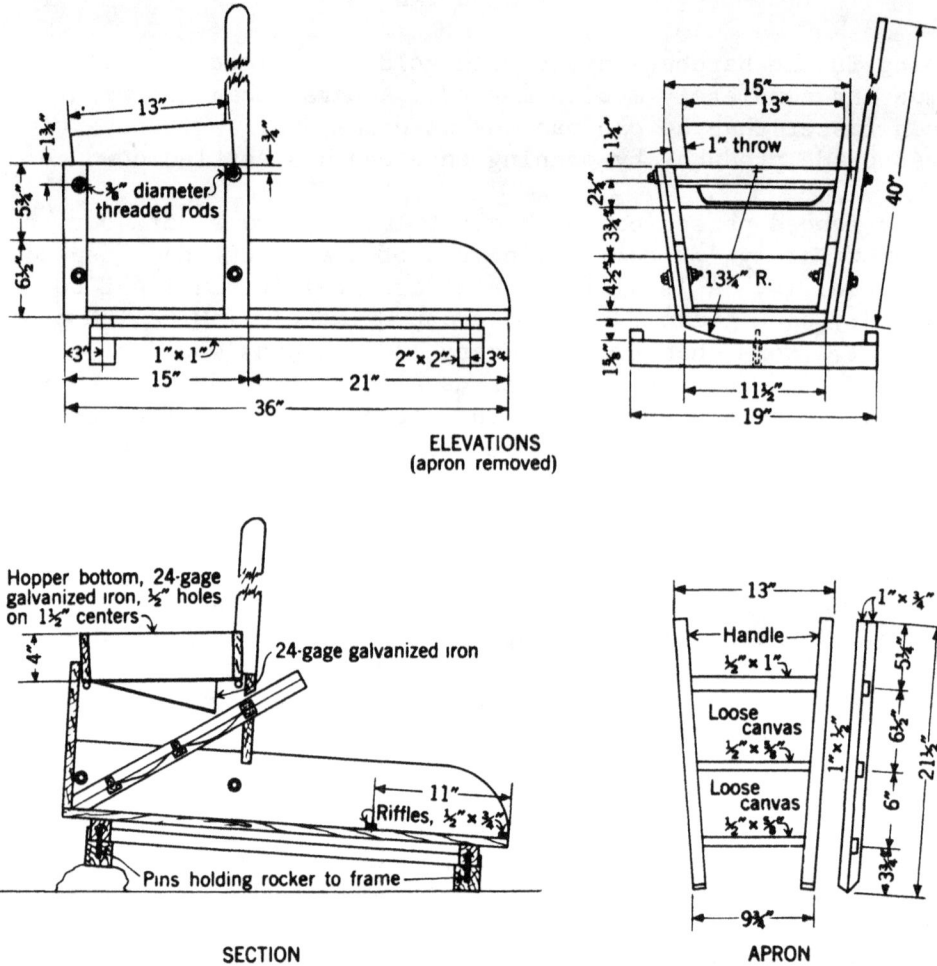

FIGURE 3. - Basic Design for a Prospector's Rocker. (Note that hopper is built to slide
back and forth, bumping the sides as unit is rocked.)

oftener when rich ground is worked and gold begins to show on the apron or in
the riffles. In cleaning up after a run, water is poured through while the
washer is gently rocked, and the top surface sand and dirt are washed away.
Then the apron is dumped into a pan. The material back of the riffles in the
sluice is taken up by a flat scoop, placed at the head of the sluice, and
washed down gently once or twice with clear water. The gold remains behind on
the boards, from which it is scraped up and put into the pan with the concentrate from the apron. The few colors left in the sluice will be caught with
the next run. The concentrate is cleaned in the pan.

Skillful manipulation of the rocker and a careful cleanup permit recovery of nearly all the gold. Violent rocking should be avoided, so that gold will not splash out of the apron or over the riffles. The sand behind the riffles should be stirred occasionally, if it shows a tendency to pack hard, to prevent loss of gold. If the gravel is very clayey it may be necessary to soak it for some hours in a tub of water before rocking it.

Where water is scarce, two small reservoirs are constructed, one in front and the other to the rear of the rocker. The reservoir at the front serves as a settling basin. The overflow drains back to the one at the rear, and the water is used over again.

The capacity of rockers may be increased by using power drives. Such a device might be rocked by an eccentric arm at the rate of approximately forty 6-inch strokes per minute. The capacity of the typical machine with two men working is 1 cubic yard per hour. Where gravel is free from clay, the capacity may be as great as 3 cubic yards per hour. The cost of the mechanized rocker and a secondhand engine for driving it is estimated at $400.

Dip-Box

The dip-box is useful where water is scarce and where an ordinary sluice cannot be used because of the terrain. It is portable and will handle about the same quantity of material as the rocker.

Construction is relatively simple. The box has a bottom of 1- by 12-inch lumber to which are nailed 1- by 6-inch sides and an end that serves as the back or head. At the other end is nailed a piece approximately 1 inch high. The bottom of the box is covered with burlap, canvas, or thin carpet to catch the gold, and over this, beginning 1 foot below the back end of the box, is laid a 1- by 3-foot strip of heavy wire screen of about 1/4-inch mesh. The fabric and screen are held in place by cleats along the sides of the box. Overall length may be 6 to 8 feet, although nearly all gold will probably collect in the first 3 feet. The box is placed so the back is about waist high; the other end is 1/2 to 1 foot lower. Material is simply dumped or shoveled into the upper end and washed by pouring water over it from a dipper, bucket, hose, or pipe until it passes through the box. The water should not be poured so hard that it washes the gold away. Larger stones (after being washed) are thrown out by hand, or a screenbox can be added to separate them. Riffles may be added to the lower section of the box if it is believed gold is being lost.

Long Tom

A long tom usually has a greater capacity than a rocker and does not require the labor of rocking. It consists essentially of a short receiving launder, an open washing box 6 to 12 feet long with the lower end a perforated plate or a screen set at an angle, and a short sluice with riffles (fig. 4). The component boxes are set on slopes ranging from 1 to 1-1/2 inches per foot. The drop between boxes aids in breaking up lumps of clay and freeing the contained gold.

A good supply of running water is required to operate a long tom successfully. The water is introduced into the receiving box with the gravel, and

26

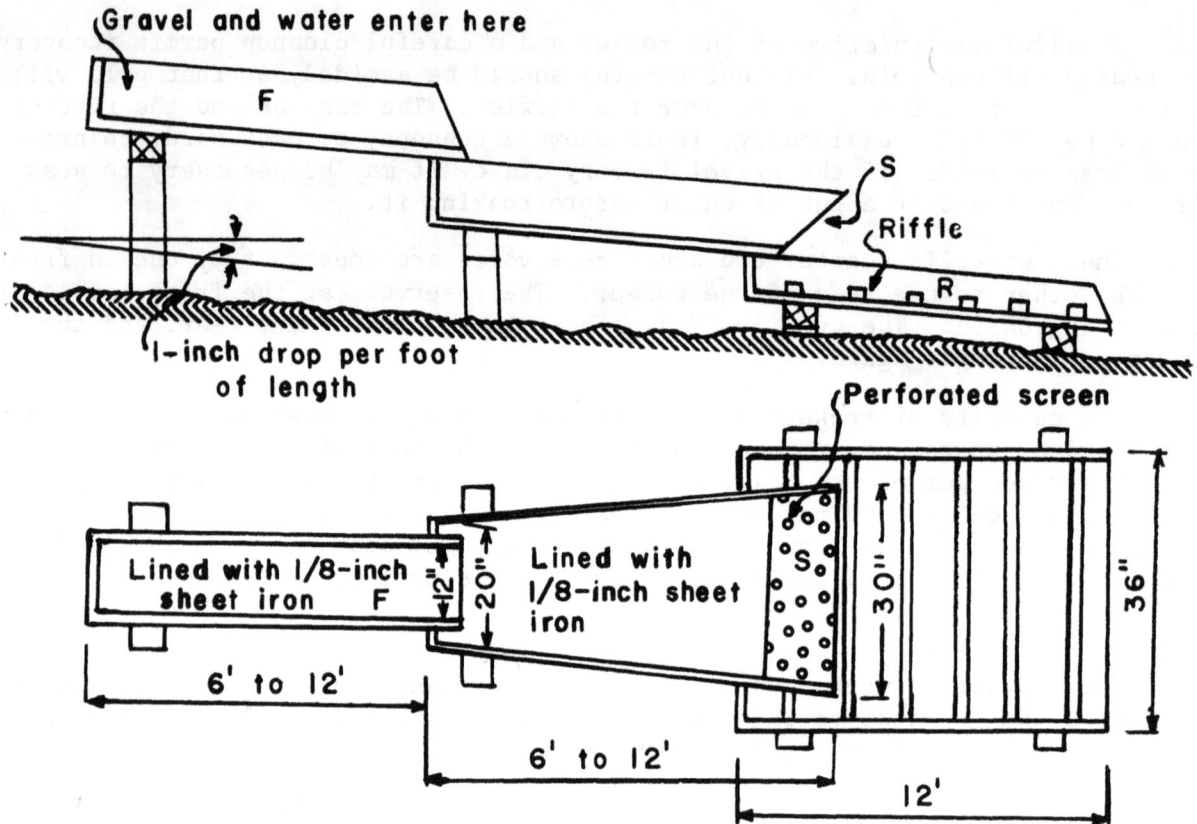

FIGURE 4. - The Long Tom.

both pass into the washing box. The sand and water pass through the screen's
1/2-inch openings and into the sluice. The oversize material is forked out.
The gold is caught by the riffles. The riffle concentrates are removed and
cleaned in a pan. Quicksilver may be used in the riffles if the gravel contains much fine gold.

The quantity of gravel that can be treated per day will vary with the
nature of the gravel, the water supply, and the number of men employed to
shovel stones into the tom and then fork them out. For example, two men, one
shoveling into the tom and one working on it, might wash 6 cubic yards of ordinary gravel, or 3 to 4 cubic yards of cemented gravel, in 10 hours.

A tom may be operated by four men--two shoveling in, one forking out
stones, and one shoveling fine tailings away. Where running water and a grade
are available, a simple sluice is generally as effective as the long tom and
requires less labor.

                              Sluice

A sluice is generally defined as an artificial channel through which flows
controlled amounts of water. In gold placering, the sluice includes sluiceboxes which collect the gold by means of various configurations of riffles,
corrugations, mats, expanded metal, or the like, which trap the heavier particles while allowing the waste to continue through. Figure 5 shows a portable

FIGURE 5. - Lightweight Sluicebox in Action. Unit made of aluminum has screen box
at the head end, riffle section in the middle, and expanded metal over
burlap at the foot. The gold pan would normally be omitted, but is
included here to check for losses.

lightweight metal sluicebox suitable for test work or a small-scale placer operation. An important part of any sluicing operation is its water supply, and where water is not plentiful, pumps, pipelines, or even dams with special headgates may be required.

Small-scale sluicing by hand methods has been called quite appropriately shoveling-into-boxes. In contrast, in ground sluicing, usually a more efficient operation, most of the excavation is accomplished by the action of water flowing openly over the materials to be mined. In either case, the materials pass through a sluice, where gold is collected behind riffles. A variation of the sluicing technique, where water is stored and released against or across the materials intermittently, is called booming.

The sluicebox in its simplest form might be a 12-foot-long plank of 1- by 12-inch pine lumber, to which sides about 10 to 12 inches high are nailed, with braces secured at several places across the top. Larger sluices can be made with battens to cover joints between boards where gold might slip out, and with braces built around the outsides of the box for greater rigidity. To provide for a series of boxes, the ends should be beveled or the units tapered so that one will slip into the other in descending order and form a tight joint. Four to eight such boxes in series would be a typical installation. Two men hand-shoveling into sluiceboxes can wash 5 to 10 times as much gravel as could be put through a rocker in a day. The slope of the sluice and the supply of water must be adjusted so that the gravel, including larger cobbles, will keep moving through the boxes and on out. Slopes of 4 to 12 inches per 12-foot box are normal, but if water is in short supply the slope may be increased. Trestles are necessary to support the boxes over excavated ground, gulleys, or swales.

Inside the boxes, various kinds of riffles may be employed, depending upon availability of material and personal preference. The riffles, which go on the bottom, are usually set crosswise in the box, but they can also be effective when placed lengthwise, the concentrates settling between them. They may be of wood, or of strap or angle iron, or a combination of the two. Straight, round poles or a pattern of square blocks or stones can serve for riffles. Rubber or plastic strips have even been used. Durability is important for prolonged operations, so wood may be armored with metal. Expanded metal, heavy wire screen, or cocoa mats make good riffles for collecting fine gold.

A common height for riffles is 1-1/2 inches; they may be placed from one-half to several inches apart. Fastening the riffles to a rack, which is then wedged into place in the box, permits their removal. A tapered shape on the cross riffle, with the thinnest edge to the bottom, tends to create an eddying action that is favorable for concentration. Another way to achieve this eddying action is to cant the riffle or even just the top of the riffle. Burlap or blanket material is commonly placed under the riffles to help in collecting fine gold. Mercury may be added to some sections of the sluice if there is much fine gold, but care must be taken to prevent escape of the mercury.

Sluice cleanups should be made at fairly regular intervals. After running clear water until the sluice is free of gravel, riffles are removed in sections starting at the upper end. With a thin stream of water, the lighter of the remaining material is washed to the sections below. The gold, heavy sands, and amalgam, if mercury has been used, are scraped up and placed in buckets. This mixture then can be panned or cleaned up in a rocker to obtain a final concentrate or amalgam.

Feeding the Sluice

It is common in a small operation, when feeding the sluice, to place a heavy screen or closely spaced bars of some sort across the section where the gravels enter, to eliminate the larger particles, which are probably barren anyway. The screen or bars (a "grizzly") should be sloped so the oversize material rolls off to the side. The size of mesh or spacing will depend upon the gradation of feed, but would generally be in the range of 1/4 to 1 inch, with 3/8 inch being a common size. In larger operations a rotating screen, or trommel, might be used. In a ground sluicing operation, possibly all materials would be run through the sluiceboxes. Provisions must be made for removing the oversize material, and, if required, stacking it away from the work area.

If the gravel contains much clay it may be desirable to use a puddling box at the head of the string of sluiceboxes. This may be any convenient size--for instance, 3 feet wide by 6 feet long, with 6- to 8-inch sides. The clayey material is shoveled into this box and broken up with a hoe or rake before being allowed to pass into the sluice. The importance of this step is that if allowed through the sluice, the unbroken clay lumps may pick up and carry away gold particles already deposited.

Usually, the shoveling-in method proceeds as follows: After the boxes are set, shoveling begins at an advantageous point. Experienced miners work out the ground in regular cuts and in an orderly fashion. Enough faces are provided so that shovelers will not interfere with one another. Provision is made to keep bedrock drained, and boulders and stumps are moved a minimum number of times. Cuts are taken of such a width and length that shoveling is made as easy as possible. The boxes are kept as low as possible so a minimum lift of gravel is necessary. At the same time an adequate slope must be maintained for the gravel to run through the boxes under the limitations of the available water. Allowance for dump room must also be provided at the tail end of the sluice. Leaks in the sluice are stopped promptly, and shoveling is done in such a manner that the sluice does not become clogged nor does water splash out. (Water in the pit hampers shoveling.)

All material of a size that will run through the sluice is shoveled in, and the oversize material is thrown to one side. Boulders from the first cut should be stacked outside the pit, on barren ground if possible. The width of a cut is usually limited to the distance a man can shovel in one operation. When shoveling from more than several feet away, it is best to set boards above and on the opposite side of the box; this increases the efficiency of the shovelers. The greatest height a man can shovel into a box is 7 to 8 feet, and above 5 or 6 feet the efficiency of the shoveler is markedly reduced. If the gravel is over 3 or 4 feet deep, it usually is excavated in benches to facilitate digging and to permit the upper layers to be raised a minimum shoveling height. Where the gravel is shallow, wheelbarrows may be used. Another way is to shovel the gravel onto a conveyor belt that discharges into a trommel, discarding the oversize material and running the undersize material through the sluice. Where two or more persons are working in the same cut, the height of succeeding benches is governed by the character of the material being dug and the distance the gravel has to be lifted.

The sluice may be maintained on the surface of unworked ground or supported on bents on the opposite side of the cut. After the first cut the boulders are thrown onto the cleaned-up bedrock. Where cuts are run on both sides of the sluice, the boxes are supported on bents as the ground underneath them is dug out. At other places the boxes may be set on bedrock and the dirt may be shoveled into the head of the sluice from short transverse cuts at the upper end of the pit. Work usually begins at the lower end of a deposit so that bedrock may be kept drained, and then proceeds across the deposit by regular cuts. The length and order of the cuts will depend upon local conditions. As heavy sands and gravel build up deposits between the riffles in the sluice, it may be necessary to stir these up to prevent packing and the consequent override of gold particles. A tined implement such as a pitchfork is often convenient for this. Larger stones that lodge in the sluiceway may be similarly removed.

Supplying Water

The quantity of water available will influence the scale of operations and the size of sluice used. A minimum flow of 15 to 20 miner's inches (170 to 225 gallons per minute) is required for a 12-inch-wide sluicebox with a steep grade. Smaller flows than this can be utilized by storing the water in some kind of reservoir and using the supply intermittently. A common practice followed where the quantity of water is limited is to use a grizzly or screen over the sluice to eliminate oversize material and thus increase the duty of the water. Reduction in the amount of material to be treated by first running it through a trommel to wash and screen out the coarse size is another effective way to lower the water requirements.

. Water usually is conducted via ditch to the sluice. However, if the ground is rich enough it may be practicable to pump water for the sluice. The feasibility of obtaining a gravity flow should first be investigated, as the expense of pumping may be more than the cost of a long ditch, when the cost is distributed over the yardage of gravel moved. A suitable number of sluiceboxes or some other removal system may be used to transport the tailings to a dumping ground away from the working area. A tailings or settling pond may be required to maintain downstream water quality.

Ground sluicing utilizes the cascading effect of water to break down the gravel; hence, the requirements for water are much greater. The chief application of ground sluicing is to streambed deposits. Pipelines, flumes, or ditches would be necessary if ground sluicing were applied to gravels higher up on banks or terraces, and the larger scale hydraulic methods would then become more favorable. If booming is to be done, a dam and reservoir are needed. The dam is usually equipped with a gate mechanism that permits either automatic or manual control and quick release of the impounded water for maximum washing effect. The water may be passed over the upper face of a gravel bank or diverted against the bottom in order to undercut and carry away the gravel as the face of the bank breaks down. All materials are channeled toward the sluice.

The natural flow of a stream can be used by diverting the current with boards or simply with piled boulders. "Shears" can be constructed of 1- or 2-inch-thick boards 12 feet long nailed to pairs of tripods so that the boards slope back from the water flow at an angle of about 60°. The tripods are built in such a way that boulders can be piled inside the base to hold them in place. A row of these shears may be used to divert the force of the water against a bank, or two rows may be used to form a flume.

The seasonal nature of stream flow in different areas must be kept in mind when planning any placer operation. State and Federal agencies can provide information on stream runoff for many of the more important streams, information which will indicate the limitations in water supply that might be expected due to seasonal changes.

Additional Methods Sometimes Used

The methods described below, particularly the surf washer, are limited in application, but interest in them revives from time to time, so they are included here. Many kinds of dry washers have been developed, some very elaborate. Most dry-washing operations have a short lifespan, owing to the erratic character if the deposits. Skindiving for gold is not new, but development of better diving equipment in recent years has stimulated interest in the method, although restricted in practice to a few select stream areas. Shaft and drift mining are also among methods used in extracting placer gold gravels, but because techniques are more related to other types of mining, discussion is not included in this report.

Dry Washer (for Desert Areas)

Dry washers have been used for many years in the Southwestern United States, where water is scarce, and especially in New Mexico where several million dollars in gold has been produced during the last century by dry washing. The Cerrillos, Golden, and Hillsboro districts are among those having produced gold by dry washing. In years when other employment is scarce such production may take place widely. In the 1930's a considerable number of men also used dry washers in Nevada, southern California, and Arizona.

If gravel is to be treated successfully by dry washing, it must be completely dry and disintegrated. For instance, after rainstorms, operations must be stopped until the ground dries out again. Even in very dry climates the gravel is slightly damp below the surface, and must be dried before it can be treated in a dry washer. Spreading the material to sun-dry or putting it through dryers adds to the cost of mining. In small-scale work, however, the gravel will dry out about as fast as it can be treated.

Dry washers are usually run by hand and have about the same capacity as rockers of corresponding size, but the work of operating the dry washer is much harder. The workers select the material they are to treat with regard to both dryness and probable gold content. It is difficult to do this on a large scale with hired labor. Plants with mechanical excavators and complex power-driven dry-washing machinery have been tried, but in the United States, at least, virtually all were commercial failures, primarily because the gravel was dug faster than the sun could dry it out. Also, in large-scale work, particularly with mechanical excavation, the cost of sizing the material is quite great. Clay and cemented gravel introduce even further difficulties.

When the gold-bearing material is completely dry and disintegrated, panning tests of the tailings should show that a good saving can be made, except perhaps with extremely fine or flaky gold. Completely disintegrated material, however, is seldom obtained. The tops of clay streaks in the gravel are likely to be richer in gold than the gravel itself. Clay or cemented gravel seldom can be broken up sufficiently by hand to free all the gold without the use of some form of pulverizer. In a dry washer all gold included in a lump of waste passes out of the machine. As water usually will break up all the gravel and separate the gold from the other material, a better saving usually can be effected with the rocker or sluiceboxes than with a dry washer.

Basically, the dry washer separates gold from sand by pulsations of air through a porous medium. The screened gravel passes down an inclined riffle box with cross riffles. The bottom of the box consists of canvas or some other fabric. Under the riffle box is a bellows, by which air in short, strong puffs is blown through the canvas. This gives a combined shaking and classifying action to the material. The gold gravitates to the canvas and is held by the riffles, while the waste passes out of the machine.

The gravel is shoveled into a box holding a few shovelfuls at the head of the washer, from which it runs by gravity through the machine. A screen with about 1/2-inch openings is used over the box. All stones over about 1 inch in diameter generally are discarded in mining. A dry washer usually is run by a small gasoline engine which saves the labor of one man. The capacity of such machines is considerably greater than that of hand-operated ones. For instance, one man working alone must fill the box, then turn a crank which runs the bellows until the gravel runs through. The process is then repeated. With two men working, one shovels and the other turns the crank. One man can treat 1/2 to 1 cubic yard per day with a hand-operated washer, where the gravel lies close to the machine.

When cleaning up, the material behind the riffles usually is dumped into a pan and washed out in water. If water is very scarce, the accumulated material from the riffles may be run through the machine a second time and then further cleaned by blowing away the lighter grains of sand in a pan.

Dry washers are usually handmade and have been built in a large number of designs and sizes. Figure 6 shows an example of one type. The bellows of the

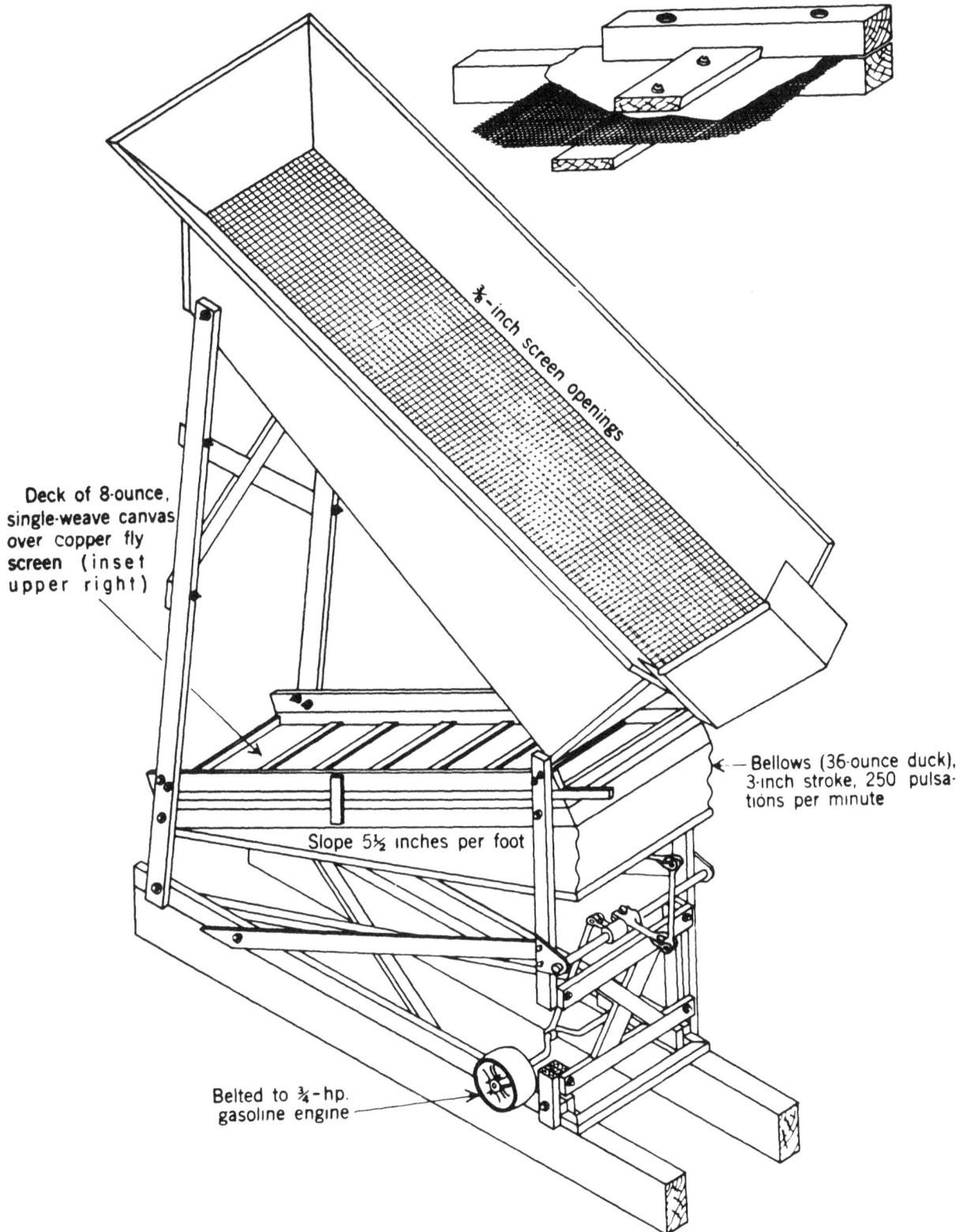

Deck of 8-ounce, single-weave canvas over copper fly screen (inset upper right)

⅜-inch screen openings

Bellows (36-ounce duck), 3-inch stroke, 250 pulsations per minute

Slope 5½ inches per foot

Belted to ¾-hp. gasoline engine

FIGURE 6. - A Dry Washer. (May be hand- or machine-powered.)

machine is made of 36-ounce duck and the bottom of the riffle box of 8-ounce,
single-weave canvas. In contrast to the single-weave canvas, silk or rayon
permits a good extraction of gold, but too much dust goes through into the
bellows. Heavier canvas is too tight for good separation. Copper-wire fly
screen is used under the canvas. The riffle box is 11 inches wide and
40 inches long and contains six riffles. The slope of the riffle box is
5-1/2 inches to the foot. (Hand-operated machines are usually much smaller
and the riffle box is set at a steeper angle than with powered machines.) The
gravel and sand are shoveled onto a screen with 3/8-inch openings at the top
of the washer. The bellows is operated at 250 pulsations per minute; the
stroke is 3 inches. The capacity of the machine is about 4/5 yard per hour,
which probably would correspond to 1-1/2 or 2 cubic yards, bank measure. (The
plus 1-inch material was previously discarded.)

In cleaning up after treating approximately 1 cubic yard in the washer,
the riffle box is lifted out and turned over on a large, flat surface, such as
a baking tin. The concentrate from the upper three riffles is first panned,
and the gold is removed. Usually both the coarse and the fine gold can be
saved here. The lower riffles may contain a few colors, but nearly all the
gold is normally caught in the upper riffles.

Surf Washer (for Beach Deposits)

Few sea-beach-type placer gold deposits have been mined successfully.
The most important producers have been in the vicinity of Nome, Alaska, but
gold is also known to occur in a few other shoreline locations of States bor-
dering the Pacific Ocean. Special techniques have been developed to utilize
the action of the surf in recovering gold from these deposits.

Surf washers are similar to long toms, but wider and shorter. They can
be used only when the surf is of proper height. They are set so the incoming
surf rushes up the sluice, washes material from the screen box or hopper, and
retreating, carries it over the riffles and plates. One man can attend to two
surf washers, and about 8 cubic yards can be handled per 10 hours.

An example of a simple surf washer is a riffled sluice 3 to 4 feet wide
and 8 to 10 feet long, set on the sand at the water's edge so that the incoming waves wash through it to the upper end, and retreat below the lower end.
The sluice is made of boards nailed to sills at either end which can be
weighed down with rocks or otherwise. The sides are 4 or 5 inches high. The
riffles in the example are made in sections of about 1- by 1-inch strips spaced
about 1 inch apart. The end sections are transverse riffles, the center section longitudinal. The box preferably is set on a grade of 8 to 10 inches per
12 feet. Best results are obtained by using mercury in the riffles. When the
surf is strong, the washer treats as much as two men can shovel, but at other
times it has to be fed very slowly.

Skindiving (Combining Recreation)

In recent years skindiving enthusiasts have taken up small-scale placering as both a hobby and a sometimes, though seldom, profitable venture.

Various kinds of apparel and equipment are used, but the investment is usually not great. Wet suits and canvas shoes are almost a necessity for entering cold mountain streams to search the streambed for pockets that might contain gold. Beginners should be equipped with a snorkel, a face mask, gloves, a weighted belt, fins, a gold pan, and a crevicing tool. More experienced divers may use the popular scuba equipment, but this calls for special knowledge to insure safety. Crevicing tools include large spoons, tire irons, crowbars, etc.--almost anything that can reach into tight places and dislodge nuggets from the stream bottom. The pan should be used to test sands from various places where gold would be expected to settle, such as the downstream sides of obstructions. Where colors in the pan indicate a favorable area of the stream, a more intense search may be made.

Mining equipment may include various combinations of pumps, miniature dredges, and riffle boxes that can be built from salvage by the operator or purchased from commercial sources. A number of manufacturers have produced special equipment for the purpose. One of the popular kinds is the jet dredge, a pipelike device made of sheet metal curved at the intake end and with a water jet entry to propel the water and gravel through the straight portion. The jet is supplied from a portable pump and in effect causes gravel and sand to be sucked into and through the pipe. A riffle box built into the end section collects the gold and other heavy particles while the rest of the material discharges. The riffle box may be enclosed so it can function while submerged. Usually, a 6- to 10-horsepower pump is adequate; the hose to the jet may be 1-1/2 to 2 inches in diameter.

Manipulating the device underwater requires skill and patience, since the riffle section must be kept nearly horizontal during the mining operation. Floating platforms are sometimes used to support equipment. In this case, riffle boxes and other units may be installed on the platform. The usual operation includes moving many large boulders to get at the trapped gold underneath or alongside. Conventional equipment such as a rocker or a sluice may be employed to carry selected material from the streambed to a shoreline site for processing. Concentrates are then panned to recover the gold.

PROBLEMS YOU SHOULD ANTICIPATE IN PLACER MINING

Besides the many problems already discussed, such as where and how to find a placer deposit, how to locate a claim, and how to sample and mine, a few special operational problems should be considered. These relate to the physical nature of placer materials and the climatic conditions under which they may be found.

Streams with steep gradients often have poorly sorted sands and gravels, meaning a wide range of size will be encountered, up to cobbles and large, irregularly shaped boulders. Other debris and tree roots may be present too. Materials that have lain in place for long periods become indurated (that is, bound up tightly with clay, or cemented sometimes almost to the point of being solid rock), which makes them exceedingly difficult to break up with water. Irregularities in the rock surface underlying placer materials become important in mining because this is the zone where the richest values usually are found. A very uneven surface can be particularly difficult to work on. In addition, there is difficulty in Alaska where ground may be frozen a large part of the year. It may be impractical for the weekend or vacation prospector to tackle placers where such adverse conditions prevail. How these problems are normally dealt with in larger operations is discussed briefly under the headings to follow.

Handling Boulders

Boulders are best left in place if it is at all possible to work around them. Sometimes, particularly in sluicing, it becomes necessary to move the boulders out of the way. A derrick operated by a hand winch or steam, gasoline, or electric power may be used for this purpose. Possibly several such derricks will be needed if many boulders are present. Boulders may be drilled with a jackhammer and blasted using dynamite, or more simply blasted with an explosive plastered onto the rock, a technique called "mudcapping." Platform skips may be swung from a derrick boom or cableway; the larger rocks are then pried out and rolled into the skip for removal. A small bucket-loader vehicle may be useful for handling boulders, provided it can operate over the type of surface exposed on the pit floor. Sections of the pit where bedrock has been cleaned up may be reserved for stacking large rocks. Future operations should be planned so repeated handling is avoided.

Trouble With Clays and Cemented Gravels

Clays and cemented gravels usually require the cutting force of the hydraulic giant for effective mining. In some nonfloating washing plants the gravel is delivered to the head of the sluice where a giant is used to break up the clay. Indurated or clayey materials are normally dredged with little difficulty, but if gravels are tightly cemented, they may best be mined by shaft or drift methods using explosives and timbering as required. This presumes they are rich enough to stand the high cost of such mining and are not exposed enough for open pit mining. Clay lumps must be broken up quite thoroughly before passing through gold-recovery equipment because of their capacity to imbed gold particles and carry the gold out with the discharge. The breaking of clays can be accomplished using the puddling box (previously described on p. 29) or with a trommel, which quickly reduces the lumps by its rotation and abrading action. Exposure of clays to air is also effective in breaking them down, although the time required may be a matter of days or weeks.

Cleaning Bedrock

Cleanup of the last remaining materials from bedrock is an important step in gold placering, and if the surface is soft, fractured, or uneven, this can be a painstaking chore. Where bedrock is soft and fractured, gold particles can be embedded as much as several feet, so it often is advisable to also excavate this kind of bedrock material for its gold content. Usually, it is best to clean the bedrock as the work progresses upstream. A final cleaning of the surface may be left until the end of the season, when there is more time to spend on this activity and when the water is short for other work.

Where bedrock is hard it must be cleaned largely by hand, and the soft
seams and cracks invariably present should be cleaned out with handtools. A
hose and small pump are almost necessities for a good cleanup. Sometimes a
separate sluicebox smaller than that used in the main operation will be
employed for handling materials from a cleanup operation.

RECOVERING YOUR GOLD AND SELLING IT

As you reach the final stage in turning arduous labors into a product,
the gold should be in either of two forms--a nearly pure concentrate or an
amalgam with mercury--depending upon whether the latter was used to implement
the collection of gold. Placer gold in its natural form is almost always
alloyed with a certain amount of silver, which decreases its fineness. The
silver, being much lower in value or price per ounce, lowers the value of the
gold by a corresponding amount. Fineness is based on a scale of 0 to 1,000.
As an example, gold 750 fine would be three-fourths gold and probably close to
one-fourth silver. The important thing is that the gold until it is refined
will be worth somewhat less than the market price for pure gold. The exception to this, of course, is specimen material that may have special value in
its natural form.

Gold in an amalgam can be heated or retorted to drive off the mercury,
leaving a gold sponge. Great care should be taken when this is done to avoid
inhalation of the mercury fumes, which are highly toxic and which can cause a
variety of ailments or even death. Small quantities of amalgam may be heated
on an iron surface, such as a shovel face, out-of-doors where the vapors will
be quickly dispersed. Preferably, a retort is used for environmental reasons
and personal safety. Mercury, which partially vaporizes at ordinary room tem-
peratures, will vaporize completely at about 675° F, so an ordinary fire or
propane burner will suffice. Small retorts are commercially available, or
they can be constructed out of a small cast iron pot with a tight-fitting
cover to which a short length of water-jacketed condenser pipe is connected.
A typical setup may have a sloping pipe 2 to 4 feet long encased in a larger
diameter pipe through which water is circulated. A coating of chalk or clay
inside the pot will prevent the gold from adhering to the iron. The pot is
heated gently at first, raising the temperature gradually until mercury stops
coming from the condenser outlet. Mercury thus recovered is ready to reuse
for amalgamation, and the spongy mass of gold can be sold. Because amalgams
are difficult to sell, it is usually best to retort your own and market the
gold.

Gold is priced and sold by the troy ounce, which should not be confused
with the better known avoirdupois ounce. A troy pound consists of 12 troy
ounces and is equivalent to 0.8229 pound avoirdupois. A button of gold that
weighed 1 pound avoirdupois would contain about 14.6 troy ounces. Normally,
gold is weighed on special troy scales so the confusion in this odd conversion
is eliminated.

When selling gold, the owner must comply with a number of requirements
laid down by the Federal Government and administered by the Department of the
Treasury. The following statement relating to gold in its natural state, gold

amalgam, and retort sponge was issued by Treasury's Office of Domestic Gold and Silver Operations in January 1969 and sums up the pertinent gold regulations then in effect:

"Gold in its natural state" is defined in the Gold Regulations as being gold recovered from natural sources, which has not been melted, smelted or refined, or otherwise treated by heating or by a chemical or electrical process. This gold may be purchased, held, sold, transported within the United States, imported or held in custody for domestic account only without a Treasury Department license under the provisions of Section 54.19 of the Regulations, regardless of the amount involved.

Gold in its natural state which has been recovered from natural sources in the United States and which has not entered into industrial or monetary use, may be exported without a license. In connection with the exportation of such gold, the exporter is required to execute and file in duplicate a certification on Form TG-34 on which information is required concerning the amount, source and description of the gold, and the consignee. Copies of this form are available from the Office of Domestic Gold and Silver Operations, Department of the Treasury, Washington, D.C. 20220. The executed form should be filed in duplicate--the original with the customs office at the port of export and the copy with the Office of Domestic Gold and Silver Operations.

Pursuant to amendments to the Gold Regulations which were effective March 18, 1968, the U.S. mints and assay offices no longer purchase gold from private sources.

Gold in its natural state and gold amalgam may be melted, smelted or refined or otherwise treated by heating or by a chemical or electrical process only pursuant to a Treasury license or without a license within the limitations contained in Section 54.19 of the Gold Regulations, as explained below.

Gold amalgam results from the addition of mercury to gold in its natural state. Gold amalgam produced from domestic sources may be dealt with in the same manner as gold in its natural state.

In addition, gold amalgam may be heated to a temperature sufficient to separate the mercury from the gold (but not to the melting temperature of gold), without a license by the person, or his duly authorized agent or employee, who recovered the gold from natural deposits in the United States or a place subject to the jurisdiction thereof. The retort sponge (amalgam cake) resulting from the heating of the amalgam may be held and transported by the person who mined or panned the gold, without a license, except that he may not hold at any one time an amount of retort sponge produced by him which exceeds in fine gold content 200 fine troy ounces.

Retort sponge produced by a miner or panner may be sold to a person holding a Treasury Department gold license authorizing the purchase of such gold, or to unlicensed persons provided that such unlicensed persons do not hold, at any one time, more than 200 fine troy ounces of gold. Persons other than the miner or panner, who acquire retort sponge, may sell the gold only to the holder of a Treasury license.

An unlicensed person may not retort gold purchased by him from miners or other persons, nor may he sell the retort sponge resulting therefrom.

Gold in melted or treated form may be sold or disposed of only by persons and concerns operating under a Treasury gold license authorizing the disposition of gold in such form.

In addition, buyers of gold may also be required in some States to hold a State license.

The dollar value received for gold will vary somewhat since the introduction of a two-level price system in 1968. Previously the U.S. Treasury had purchased gold for $35 an ounce, less a small charge for service and melting. Currently, one price exists for official monetary transactions at $35 an ounce while another exists for private transactions based on open-market demands. The price paid to the private seller of gold will depend not only upon the fineness and purity of the gold but on day-to-day market fluctuations. Financial pages of most newspapers and industry trade journals should be consulted for latest price quotations.

Finding a buyer for small lots of gold is not as easy as one might expect. Normally, an assay is necessary to establish the purity, and a sample must be taken in such a way that it is representative of the whole amount. A typical smelter accepting gold (American Smelting & Refining Co. at Selby, Calif.)[5] quoted a refining charge of $30 per lot (1970) for such services plus a small melting charge that varied according to the quantity (the smaller the quantity, the greater the rate per ounce; for 1,000 ounces this charge was $6). Also, there were reductions from quoted market prices for other reasons such as impurities or special handling. When a sale is completed, the seller receives a settlement sheet which specifies gold and silver content, sales value, and the charges, and a check for the amount due.

A producer may prefer to sell to a dealer, of whom there are only a few that will purchase in small lots. Names of prospective buyers can usually be found in the classified section of the telephone book for larger cities under Gold and Silver Buyers. Such a buyer may require sampling and assay, adding the appropriate charges, or may buy simply on inspection, although discounting his offer to allow for errors in estimation. Since he will often end up selling to a smelter, his price must be low enough to permit him a profit after

---

[5]Reference to specific company names is made for identification only and does not imply endorsement by the Bureau of Mines.

covering the smelter charges and handling costs. A limited market exists in selling directly to a collector or jeweler, either for speculation on a future price increase or for use in special jewelry that requires crude gold.

After receiving payment, or after placing your gold in a display case or strong box, hopefully you will feel the effort was all worthwhile, even if your venture would not be considered a financial success. Perhaps because of the exercise and fresh air your health will be better than when you started.

SELECTED BIBLIOGRAPHY WITH NOTATIONS[6]

1. Adams, G. I. Gold Deposits of Alabama and Occurrences of Copper, Pyrite, Arsenic, and Tin. Alabama Geol. Survey Bull. 40, 1930, 91 pp. Placers and lode mines described by counties, districts, and individual properties; production statistics 1830-1916.

2. Averill, C. V. Placer Mining for Gold in California. Calif. Div. Mines Bull. 135, 1946, 377 pp. One of the best reports published specifically on California placer gold. Discussion of geology of placers, mining methods from small-scale operations through dredging, examples of operations by county and company; includes maps.

3. Bryson, H. J. Gold Deposits in North Carolina. North Carolina Dept. of Cons. and Dev. Bull. 38, 1936, 157 pp. Mostly about gold quartz vein deposits; contains brief summary on placers, particularly the saprolite type (gold with host rock decomposed in place).

4. California Division of Mines and Geology. Legal Guide for California Prospectors and Miners. 1962, 128 pp. Particularly written for California but contains much information of general value to mineral locators in other States.

5. Clark, W. B. Skin Diving for Gold in California. Calif. Div. of Mines and Geol. Min. Inf. Serv., v. 13, June 1960, 8 pp.

6. _____. Gold Districts of California. Calif. Div. of Mines and Geol. Bull. 193, 1970, 186 pp. Provides information and references on history, geology, and ore deposits by regions and districts.

7. Gardner, E. D., and P. T. Allsman. Power Shovel and Dragline Placer Mining. BuMines Inf. Circ. 7013, 1938, 68 pp. Many tables listing examples of operations.

8. Gardner, E. D., and C. H. Johnson. Placer Mining in the Western United States: Part I. General Information, Hand Shoveling and Ground Sluicing. BuMines Inf. Circ. 6786, 1934, 74 pp.; Part II. Hydraulicking, Treatment of Placer Concentrates, and Marketing of Gold. BuMines Inf. Circ. 6787, 1934, 89 pp.; Part III. Dredging and Other Forms of Mechanical Handling of Gravel, and Drift Mining. BuMines Inf. Circ. 6788, 1935, 82 pp. One of the most complete reviews of the overall subject of placer mining; contains many specific examples of operations.

9. Haley, C. S. Gold Placers of California. Calif. State Min. Bur. Bull. 92, 1923, 167 pp. Review of placer mining methods; productive rivers, creeks, and areas in California.

10. Jackson, C. F., and J. B. Knaebel. Small-Scale Placer-Mining Methods. BuMines Inf. Circ. 6611, 1932, 17 pp. Contains State maps showing locations of placer mining districts; discusses simple placer methods.

---

[6]Many of these reports are out-of-print and may only be available now through libraries.

42

11. Janin, C. Recent Progress in the Thawing of Frozen Gravel in Placer
     Mining. BuMines Tech. Paper 309, 1922, 34 pp. Examples and opera-
     tional data.

12. Johnson, F. W., and C. F. Jackson. Federal Placer-Mining Laws and Regu-
     lations; Small-Scale Placer-Mining Methods. BuMines Tech. Paper 591,
     1938, 49 pp.

13. Jones, S. P. Second Report on the Gold Deposits of Georgia. Georgia
     Geol. Survey Bull. 19, 1909, 283 pp. Lode and placer mines described
     together; many details about individual properties and past operations.

14. Koschmann, A. H., and M. H. Bergendahl. Principal Gold-Producing Dis-
     tricts of the United States. U.S. Geol. Survey Prof. Paper 610, 1968,
     283 pp. One of the most complete reviews of gold mines and districts
     published to date; includes locations of placer areas and production
     information; bibliography of geological reports.

15. Lyden, C. J. The Gold Placers of Montana. Montana Bur. of Mines and
     Geol. Mem. 26, 1948, 151 pp. Many details of specific operations,
     described by county and river or creek; good series of location maps.

16. Oregon Department of Geology and Mineral Industries (staff). Oregon's
     Gold Placers. Misc. Paper No. 5, 1954, 14 pp. Briefly describes past
     gold placer activities by stream drainages in southwestern and north-
     eastern Oregon.

17. Pardee, J. T., and C. F. Park, Jr. Gold Deposits of the Southern
     Piedmont. Geol. Survey Prof. Paper 213, 1948, 156 pp. Relatively
     technical, but one of the best reviews available on gold along the
     eastern flank of the Appalachian Mountains from Virginia to east-
     central Alabama; contains detailed lists of gold localities.

18. Peele, Robert. Mining Engineer's Handbook. John Wiley & Sons, Inc.,
     New York, 1947, 2 volumes. Technical and authoritative; divided into
     sections covering all phases of mining.

19. Purington, C. W. Methods and Costs of Gravel and Placer Mining in Alaska.
     U.S. Geol. Survey Bull. 263, 1905, 473 pp. Written during period of
     great activity in Alaskan goldfields; information on hydraulic mining,
     special problems with frozen ground, and description of early dredging.

20. Romanowitz, C. M., H. J. Bennett, and W. L. Dare. Gold Placer Mining.
     Placer Evaluation and Dredge Selection. BuMines Inf. Circ. 8462, 1970,
     56 pp. Covers dredging from "doodle-bug" to large bucket-line and off-
     shore operations, things to consider in exploration and evaluation of
     deposits, typical operating costs.

21. Staley, W. W. Gold in Idaho. Idaho Bur. of Mines and Geol. Pamph. 68,
     1946, 32 pp. Gives placer production by county; includes district maps
     showing locations of gold veins, mines, and prospects.

22. Thomas, B. I., D. J. Cook, E. Wolff, and W. H. Kerns. Placer Mining in Alaska: Methods and Costs at Operations Using Hydraulic and Mechanical Excavation Equipment With Nonfloating Washing Plants. BuMines Inf. Circ. 7926, 1959, 34 pp. Some typical operations in recent years. Describes "sluiceplate" mining.

23. U.S. Bureau of Land Management. Regulations Pertaining to Mining Claims Under the General Mining Laws of 1872, Multiple Use, and Special Disposal Provisions. Circ. 2149, 1964, 45 pp. Covers regulations on placer location and patenting.

24. U.S. Bureau of Mines. Production Potential of Known Gold Deposits in the United States. BuMines Inf. Circ. 8331, 1967, 24 pp.

25. U.S. Department of the Air Force. Search and Rescue Survival. Air Force Manual, AFM 64-5, August 1969, 136 pp. Contains a brief, concise section on first aid and has many suggestions on how to survive in a variety of situations.

26. U.S. Department of the Treasury. Gold Regulations. (Revised May 1, 1969), 17 pp. States regulations regarding possession, transport, foreign trade, and sale of gold in all forms.

27. Vandenburg, W. O. Placer Mining in Nevada. Univ. of Nev. Bull., v. 30, No. 4, May 15, 1936, 180 pp. Particularly good discussion of placer mining where water is short. Nevada deposits discussed by county and district.

28. Vanderwilt, J. W., and others. Minerals Resources of Colorado. Denver, Colo., Min. Resources Board, 1947, 547 pp. Contains good maps showing locations of Colorado placers.

29. von Bernewitz, M. W. Treatment and Sale of Black Sands. BuMines Inf. Circ. 7000, 1938, 21 pp. Concentration, amalgamation, and extraction of gold from heavy sands.

30. Wells, E. H., and T. P. Wooten. Gold Mining and Gold Deposits in New Mexico. New Mexico Bur. of Mines and Min. Resources Circ. 5 (revised ed.), 1940 (report reissued in 1957), 24 pp. Limited information on placers; deposits described by counties and districts.

31. Wells, J. H. Placer Examination: Principles and Practice. Bur. of Land Management Tech. Bull. 4, 1969, 155 pp. Excellent technical report on sampling procedure and estimation of gold values in placer deposits; contains glossary of terms used in placer mining.

32. Wilson, F. D. Arizona Gold Placers, Part 1 of Arizona Gold Placers and Placering. Arizona Bur. of Mines Bull. 160, Min. Tech. Series 45, 1952, pp. 11-86. Placers described in good detail by county and district; many individual operations cited.

33. Wimmler, N. L. Placer-Mining Methods and Costs in Alaska. BuMines Bull. 259, 1927, 236 pp. Useful information on all methods of placer mining; emphasis on Alaskan problems with seasonal activities.